T0181715

Development Practice in Eastern and Southern Africa

Tshilidzi Madzivhandila · Sepo Hachigonta ·
Joseph Francis · Joseph Kamuzhanje ·
Oluwatoyin Dare Kolawole · Shirley DeWolf
Editors

Development Practice in Eastern and Southern Africa

Lived Experiences from the Trenches

Editors
Tshilidzi Madzivhandila
Food, Agriculture and Natural Resources
Policy Analysis Network
Pretoria, Gauteng, South Africa

Joseph Francis
Institute for Rural Development
University of Venda
Thohoyandou, Limpopo, South Africa

Oluwatoyin Dare Kolawole
Okavango Research Institute
University of Botswana
Maun, Botswana

Sepo Hachigonta
National Research Foundation
Pretoria, Gauteng, South Africa

Joseph Kamuzhanje
Coopers Zimbabwe Pvt Ltd
Harare, Zimbabwe

Shirley DeWolf
Independent Development Expert
Mutare, Zimbabwe

ISBN 978-3-030-91133-1 ISBN 978-3-030-91131-7 (eBook)
https://doi.org/10.1007/978-3-030-91131-7

This Springer imprint is published by the registered company Springer Nature Switzerland AG
The registered company address is: Gewerbestrasse 11, 6330 Cham, Switzerland

Preface

With a population estimated to be more than a billion, Africa faces numerous political and socio-economic challenges, extreme poverty and underdevelopment being the most prominent. Its citizens are losing hope on how to improve their socio-political-economic conditions. Development programmes and similar interventions continue to fail to enhance the quality of life of most of its population. The COVID-19 pandemic has exacerbated the grave situation. In fact, the pandemic has exposed Africa's weaknesses and lack of investment in key infrastructure, systems and innovations including the development of vaccines and safety net programs to protect society from impact of economic shocks, natural disasters and others. This is despite the continent's long history of vulnerability to extreme climate variability and acute disease burdens such as Ebola and HIV and AIDS. All things considered, the daunting development challenges facing Africa can only be addressed through innovative, inward looking and radical approaches that specifically underscore the continent's contextual and unique socio political economic and cultural landscapes.

The idea behind this book was conceived in 2015 when a small group of academics, development experts and policy makers came together to deliberate on how to document and share their field-based experiences in southern Africa. The discussions on the initiative culminated in a workshop that the Institute of Rural Development (IRD) at the University of Venda in Thohoyandou, South Africa, hosted in April 2019. Thus, this book is a compilation of real-life experiences and lessons that academics and other professionals working on development issues in eastern and southern Africa learnt. It is a rich collection of reflective and reflexive stories that seasoned academics, leaders of institutions and journalists who have shaped Africa's development agenda and direction penned. Specifically, the book was conceived to fulfil two major purposes. First, it is meant to serve as a teaching and learning resource for undergraduate and postgraduate students undergoing training in various development subjects in colleges and universities. Second, the stories in the book are compiled to influence policy development and implementation in southern Africa and beyond. The stories are drawn from the experiences of development practitioners and researchers in Botswana, Eswatini, Kenya, Somalia, South Africa, Zambia and Zimbabwe. This means that the array of contributors' viewpoints transcend disciplinary, national and regional boundaries, which attests to the considerable promise that the book might hold in terms of its relevance and impact on societal transformation.

The book should be regarded as a veritable platform for sharing development practitioners' personal experiences, which are rarely captured in traditional scholarly publications or project reports. Imbued with multidisciplinary perspectives, the stories in the book constitute a rich collection of experiences narrated with real-life illustrations that other development practitioners, researchers and stakeholders will find useful in their own work.

As is the case with similar projects of this nature, realising the objective of this book was fraught with numerous challenges. Securing contributors from eastern and southern African countries was an uphill task. While receiving contributions within set timelines was problematic, ensuring that the individual and overall products were of good quality to include in a standard book proved to be an even bigger challenge.

It is worth restating that development practitioners work in many diverse environments where they coordinate projects, conduct scientific research and are involved in community-engaged work. For the work to be executed effectively, frontline workers should have specialised knowledge, skills, competences, attitudes and values that enable them to shape development agenda and initiatives. Through this book, we have created a platform for sharing lived experiences of the practitioners with those interested in pursuing a career in development-related issues in Africa. Therefore, we are convinced that the book will strengthen practical development practice through building the capacities of young and emerging leaders. Finally, we hope that the book will provide the much-needed ingredients for informed institutional and public policy making.

Pretoria, South Africa — Tshilidzi Madzivhandila
Pretoria, South Africa — Sepo Hachigonta
Thohoyandou, South Africa — Joseph Francis
Harare, Zimbabwe — Joseph Kamuzhanje
Maun, Botswana — Oluwatoyin Dare Kolawole
Mutare, Zimbabwe — Shirley DeWolf

Acknowledgements The editors wish to thank the authors who contributed to this book project. Special gratitude is extended to the participants in the workshops held in 2015 and 2019 for giving birth to this book. The Institute for Rural Development (IRD), Food, Agriculture and Natural Resources Policy Analysis Network (FANRPAN) and the National Research Foundation (NRF) of South Africa jointly coordinated the development of the book.

Introduction

Background

The complexity of socio-economic challenges witnessed in contemporary societies necessitates the adoption of multifaceted approaches anchored on a sound understanding of local systems and their related dynamics. While the sustainable development goals (SDGs) and Africa Union Agenda 2063 (African Union, 2019) provide the frameworks for achieving a better and more sustainable future for all, the instruments may not fully realise their goals if they are not adequately linked to tangible programmes on the ground.

Development is holistic in nature, and it transcends scope that goes beyond mere economic growth. Indeed, the concept is very complex, and its meaning is subject to how individuals view or perceive it. Its various dimensions can only be thoroughly comprehended within the context of people's yearnings and aspirations. The complexity of the subject warrants the need for those working in the field to acquire sound expertise and analytical tools for understanding the workings and machinations of any social systems and their dynamics. While theorising development may be necessary to help shape experts' thinking on the subject, it is not in any way sufficient on its own. Both theory and praxis should complement each other to achieve human growth and development. The rate at which alterations now occur in all spheres (including governance, political economy, climate and public health, among others) at both local and international scales is a testament to the fluidity and complexity of development itself. Put differently, the plethora of development challenges, which currently confronts humanity, is an avenue for pessimism (Chambers, 2005). Nonetheless, optimism must transcend pessimism only if development actors are determined to not allow the situation to go from bad to worse. If the actors become too pessimistic and allow persistent challenges that confront them on a daily basis to blur their visions, the situation can worsen to the extent of becoming *laissez faire* in their dispositions towards bringing about "good" change. To combat complacency from becoming too entrenched, a mixture of pessimism and optimism is needed. Borrowing from the Sachs' (1992) pessimism and Thuvessons' (1995) optimism about development, Chambers' (2005: p. 186) rejoinder affirms that "[a] balanced view has to recognise renewals and continuities in the landscape as well as ruins and rubble, and other trees as well as new sprouts only if we are to achieve success in the business". To consistently dwell on the failures and errors associated with development and overlook, the progress made is to invite the actor to be overly complacent and do nothing to face the challenge headlong.

Understanding how to address wicked problems goes beyond the traditional approach commonly used in development practice. This calls for going the extra mile to understand the rules of engagement and ensuring that they are properly applied in real-life situations. Ultimately, the overarching goal is to collaborate to achieve human progress in its many facets. Development is meaningful when it results in positive change in society and enhances people's lives, assets and capabilities. The multidimensional nature of development thus warrants viewing the concept through an eclectic approach, which makes it possible to fully understand issues to address (Kolawole, 2010). As Seers (1969) observes, development cannot occur within any context if the problems of unemployment, inequality and poverty remain inadequately addressed. Many more intertwining issues hamper development efforts. They include conflicts, climate change, famine, disease outbreaks, bad leadership, poor governance and corporate irresponsibility, among others.

The right and access to food, health services, education and water partly constitute the bedrock of sustainable development which is anchored on the idea that current human needs are met without compromising the ability of future generations to meet their own needs. It is worth acknowledging that African countries have made some progress towards addressing the multiple challenges that the continent faces. However, sustainable solutions remain elusive due to a myriad of issues that are mainly associated with the deficit in Africa's polity and governance. Among others, accountability and transparency, transformative leadership, commitment to long-term planning, adoption and implementation of goal-oriented approaches that consolidate the efforts and resources of multiple sectors are required to successfully achieve sustainable development on the continent. One major avenue through which this goal can be attained is the creation of the platforms for development practitioners and academics to share and deliberate on their experiences. Ultimately, such intellectual engagements and thought leadership must inform processes, programmes and policies designed to yield sustained and sustainable solutions to the multiple challenges constituting a barrier to the continent's ability to improve the livelihoods and well-being of its citizens. This book is one such strategy because it has the potential to close the gap.

This book is a product of rigorous debates among academics and development practitioners on how rural and urban development practice can be likened to a war in which the "fighters" wage battles and wars from the "trenches". It thus serves as a platform for sharing the failures and successes witnessed during the time development practitioners spent in the trenches.

Book Structure

This book is the first edition of a compilation of stories that emanated from academics and professionals working within the context of development practice in eastern and southern Africa. *Development Practice in Eastern and Southern Africa* is partitioned into four sections with diverse shared experiences that need to be conveyed to its readership.

Following this introductory chapter, Part I is devoted to the lived experiences drawn from development-oriented research in eastern and southern Africa. The chapters in this section outline how development researchers navigated multiple realities, challenges and opportunities including lessons learnt in the field. The experiences drawn from the Okavango Delta of Botswana and some parts of South Africa and Zimbabwe come to bear in this section.

Personal stories and experiences from community development practice are presented in Part II. Authors narrate and reflect on their involvement in development programmes that supported small-scale farmers. Reflections on crossborder issues add unique flavour to the collection of contributions.

Part III outlines the dynamics associated with culture, skills development, gender issues and science communication in development practice. Authors' accounts from field research and development endeavours provide interesting and rich resources. This section also lays bare some intriguing socio-cultural issues on gender and power dynamics in rural communities.

Part IV is rich in individuals' "lived" experiences that relate to career progression in rural development programmes. Among them are the stories that highlight the importance of mentorship, integrity, humility, team and hardwork in successful navigation of career paths within development programming.

The concluding section of this first volume is a synthesis that distils cross-cutting issues in the narratives. Authors' challenges and opportunities from development practice are discussed within the context of key policy issues that might provide a roadmap for enhancing development work, especially in eastern and southern Africa.

Tshilidzi Madzivhandila
Sepo Hachigonta
Joseph Francis
Joseph Kamuzhanje
Oluwatoyin Dare Kolawole
Shirley DeWolf

References

Africa Union (2019). Africa Union Agenda 2063: The Africa We Want. *African Union*. Online document: https://au.int/en/agenda2063/overview. Accessed 5 October 2021.
Chambers, R. (2005). *Ideas for development* (pp. 1–213). London: Earthscan.
Kolawole, O. D. (2010). Inter-disciplinarity, development studies, and development practice, *Development in Practice*, *20*(2), 227–239. https://doi.org/10.1080/09614520903564223

Sachs, W. (1992). *The development dictionary: A guide to knowledge as power*. London: Zed Books.
Seers, D. (1969). The meaning of development. *International Development Review, 11*(4), 3–4.
Thuvesson, D. (1995). Forest trees and people newsletter. *Editorial*, pp. 26–27.

Contents

Editors and Contributors

About the Editors

Tshilidzi Madzivhandila is CEO of the Food, Agriculture and Natural Resources Policy Analysis Network's (FANRPAN), in Pretoria, South Africa. He is an expert in public policies and programmes evaluation having studied at the School of Business, Economics and Public Policy, at the University of the New England (Australia).

Sepo Hachigonta is Director of Strategic Partnerships within the Strategy, Planning and Partnerships business unit at the National Research Foundation (NRF), in Pretoria, South Africa. His research interests are in transdisciplinary fields, food systems as well as Africa's science, technology and innovation policy landscape.

Joseph Francis is Professor and Director of the Institute for Rural Development at the University of Venda in South Africa. He has worked as a university academic for more than 23 years and has extensively engaged with grassroots communities in various countries in southern Africa. His expertise encompasses local governance, public participation, local economic development, innovation for development and poverty studies.

Joseph Kamuzhanje is Professor and Director and Lead Consultant for the Perch Inc Development Consultancy Services, Zimbabwe. He is an experienced rural and urban planner, development practitioner and social scientist with over 25 years of experience in development.

Oluwatoyin Dare Kolawole is Professor of Rural Development at the Okavango Research Institute (ORI), University of Botswana (UB), Botswana. He is Adjunct Faculty at the Eastern University based in Pennsylvania, USA, and was UC Visiting Canterbury Fellow in 2014. He works at the interface of science, policy and agriculture in Africa.

Shirley DeWolf is a pastor, community activist, mentor and retired university lecturer. She has extensive experience in rural development planning and implementation, refugee support, human resources development.

Contributors

Steve Kemp University of Edinburgh, Edinburgh, United Kingdom

Alois Sibaningi Baleni Society, Work and Politics Institute/ African Center for Migration and Society, Wits University, Johannesburg, South Africa

Lin Cassidy Okavango Research Institute, University of Botswana, Maun, Botswana

Petronella Chaminuka Agricultural Research Council, Hatfield, Pretoria, South Africa

Ephraim Chifamba Department of Rural and Urban Planning, Great Zimbabwe University, Masvingo, Zimbabwe

Joseph Francis University of Venda, Institute for Rural Development, Thohoyandou, South Africa

Emaculate Ingwani University of Venda, Thohoyandou, South Africa

Hlekani Muchazotida Kabiti Risk and Vulnerability Science Centre, Walter Sisulu University, NMD Campus, Mthatha, South Africa

Joseph Kamuzhanje Coopers Zimbabwe Pvt Ltd, Harare, Zimbabwe

Oluwatoyin Dare Kolawole Okavango Research Institute, University of Botswana, Maun, Botswana

Tshilidzi Madzivhandila Pretoria, South Africa

Langtone Maunganidze Midlands State University, Gweru, Zimbabwe

Barbara Ngwenya Okavango Research Institute, University of Botswana, Maun, Botswana

Phathisiwe Ngwenya Bulawayo, Zimbabwe

Pertina Nyamukondiwa University of Venda, Thohoyandou, South Africa

Idowu Kolawole Odubote Zambia Academy of Sciences, Lusaka, Zambia

Mantoe Phakathi University of South Africa, Pretoria, South Africa

Mosarwa Segwabe Gaborone, Botswana

Simba Sibanda Pretoria, South Africa

Sue Walker Agrometeorology, Agricultural Research Council—Soil, Climate and Water, Pretoria, South Africa;
Department of Soil, Crop and Climate Sciences, University of the Free State, Bloemfontein, South Africa

Mafuta Wonder The University of Venda, Thohoyandou, South Africa

Part I
Navigating Research in Development Practices

Chapter 1
Navigating Multiple Identities and Vulnerabilities when Conducting Fieldwork

Petronella Chaminuka

Introduction

During the last 25 years, I have worked on several community projects in remote areas that have compelled me to adopt multiple identities. Outside the field, I engaged fully with my international project partners as a researcher on 'equal grounds'. However, during field visits in South Africa, where I have carried out most of my professional work, I have often found myself assuming the roles of a researcher, research subject, advocate, protector and in some instances, a sacrificial lamb. In this chapter, I present, some of my experiences in my journey of developmental practice.

Navigating Culture Dynamics

As part of the process of community entry to introduce a project, it is common for researchers to appear before one or more traditional/community leaders, who in most instances are male. As is customary in some places, women must dress in an appropriate way, which often includes having to cover one's shoulders and hair.

On two occasions, I was part of an international research team that introduced a project to traditional leaders in the Limpopo Province of South Africa. Our research team comprised both male and female local and international researchers. During field trips, my professional portrayal from locals differed from that of my fellow female, white Global north researchers. Twice, I was denied entry into meetings without putting on head gear and a scarf to cover my shoulders. What was puzzling for me was that European female researchers in the team did not have to comply with this requirement for them to be allowed to participate in the meetings. Both times,

---

P. Chaminuka (✉)
Agricultural Research Council, 1134 Park Street, Hatfield, Pretoria 0083, South Africa

T. Madzivhandila et al. (eds.), *Development Practice in Eastern and Southern Africa*,
https://doi.org/10.1007/978-3-030-91131-7_1

sympathetic ladies working in the offices of traditional leaders came to my rescue by lending me their own head gear and scarfs.

During one meeting, we were informed that women had to kneel when greeting traditional leaders. Although the host did not necessarily assert this, my upbringing and familiarity with this tradition could not allow me to remain seated when greeting elders. During most engagements with the traditional leaders, I frequently alternated my personality between that of a researcher and a culturally sensitive and meek African girl child.

A key principle when conducting research in rural areas is having a listening and learning mindset and not having a "know it all attitude". Assuming a curious disposition is advisable, which entails always asking so that the locals tell their story without any bias. I was raised in rural areas and have a solid understanding of life in these setups. Quite often, I cringed at some of the questions that my European co-researchers asked when we conducted research in the field.

One day, I found myself listening patiently to my colleague quizzing a young woman. She was asking her why she carried a bucket of water on her head. The inquiry was about the level of discomfort associated with that, in addition to wanting to know whether the bucket would not fall and how she had learnt the art. Seeing the puzzled look on the face of the young woman being quizzed, I had to whisper to my colleague that I would provide answers at a later stage. Here I was volunteering myself as a research subject. This incident and several others later helped me understand why Global north researchers in our group who had spent so much time in the field collected much less data compared to their African counterparts. A huge gulf in understanding and conducting community-based research existed between African and Global north researchers. In fact, we often joked later with other African researchers about how our counterparts weaved their way through seemingly impressive stories on livelihoods in rural Africa based on just a few photos and information that were in most instances obvious to us.

Another experience of note was when I embarked on my PhD studies. Our project coordinator told us that one major skill he developed during his fieldwork experience was the art of 'waiting'. In rural areas, it is common for meetings to start several hours later than scheduled times. Locals who usually come early sometimes do not mind waiting until engagements start. After years of fieldwork experience and mastering a better understanding of the local context, one does not get surprised or irritated by the long delays. However, within the context of travelling with international partners, there is the underlying pressure to defend the delays. In some cases, we even apologise for such delays to our non-African research partners. Another view I have come across over the years is the notion that people come to meetings because food is provided. Some researchers even still believe that providing snacks or lunch after community meetings or workshops is unethical and might influence the participants' viewpoints. If this is the view, is it not ironic that we rarely hold meetings in our offices without coffee or lunch being served? Having shared all this, let me shift to specifically focus on the following: household surveys for HIV (human immunodeficiency virus)/AIDS (acquired immunodeficiency syndrome) studies; safety concerns and doing household surveys the "appropriate way".

Household Surveys for HIV/AIDS Studies

In 2005, while working on a study on the impacts of HIV/AIDS on agriculture, I deployed a team of enumerators one early morning to visit a village near my university in Limpopo Province. We were conducting a household survey. I thought I had trained my research assistants adequately to play their expected roles. We had pre-tested and pilot-tested our questionnaire after ethical clearance had been secured. Based on a list of households that we had randomly sampled using the lists that the local Chief had provided, I deployed the research assistants to conduct the interviews. We planned to regroup after interviewing the first set of households. As agreed, after about two hours almost everyone had reported back to our designated meeting point, namely the local shopping centre. However, one female research assistant did not turn up. We waited for about 30 min. As we wondered what was going on, I received a message on my phone requesting me to go to a household where she was conducting an interview. I got to the homestead and found her sitting with an elderly woman inside a brick and tile house. Both of them seemed to have been crying.

It turned out that our random sampling approach had landed us in one of the households worst affected by the HIV/AIDS pandemic in the village. The elderly lady was a mother to three daughters and a son who all died of AIDS within a space of two and a half years. We were conducting the interview about three months after she had buried her last daughter. One of our questions inquired about household members who had fallen sick and died from HIV/AIDS related illness in the last 24 months. This is the question that triggered the very painful memories of events that the elderly lady had experienced, leading to the crying. Even my research assistant had been overwhelmed by the emotion that prevailed. What made this experience worse was that the old lady explained that she was being labelled a witch and members of the village were ostracising her. They accused her of having bewitched and killed her own children. Her situation was so dire and emotional that we found ourselves spending the day consoling her. We decided to prepare and share a meal with her. Here I was, for the first time, faced with this type of predicament with a research assistant and myself.

At the end of that day, we found ourselves emptying our wallets as we shared with her the little cash we had. As we left the homestead, we felt helpless and regretted the unintended consequences of our search for data. My sense of helplessness was magnified by the fact that because I could not speak the local language, I failed to counsel or console the old lady to my satisfaction. That experience marked the last time I ever conducted an HIV/AIDS study without seeking the guidance of community development workers regarding the suitability of sampled households to interview, especially in areas where I was not conversant with the local language. This experience also highlights the need to teach and prepare our students and researchers for such extreme events.

Vulnerability of Female Researchers when Conducting Household Surveys

In 2007, I was conducting research on livestock production systems in one village of Limpopo Province, South Africa through household surveys, focus group discussions and key informant interviewing. To access the farmers, I was invited to a cattle dipping session at 5am on a Thursday morning. Extension and advisory personnel operating in the area were going to introduce me to the local farmers I would eventually have to work with. This was an exciting opportunity for me.

On the day, I woke up early and left the village lodge where I was staying to make way to the dip tank. As expected, out of the estimated 35 farmers at the dip tank, the majority were men. There were two elderly women who sat some distance away from where men were handling cattle. The old ladies informed me that their children were all working in cities, and thus were left to manage the cattle in the villages.

I walked up to the dip tank and greeted the extension worker who had invited me. He introduced me to the many farmers who were there. I secured appointments to interview some of them as key informants. Among them was a man who informed that he was free to be interviewed after the dipping session. He duly invited me to walk with him to his homestead for the interview. We got to the homestead. He invited me into the house. I learnt later that he was unmarried and lived alone. As we sat in the house, he invited me to make some tea for both of us if I wished. I got worried when our discussion veered off course. He questioned why such a "young beautiful woman" was walking alone around the area. That discussion made me feel vulnerable. It dawned on me that in my excitement to interview him, I had overlooked my own personal safety and security. Thus, I had to make a hasty retreat. I lied to him that I had forgotten the guideline for questions in the lodge, had to collect it and return to conduct the interview. My instincts were confirmed as later, the same man made more direct inappropriate advances to me.

Reflecting on the incident above, I found it ironic that as a research methods lecturer, I always emphasized that my students should visit and interview respondents in their own homes. From my experience of working in rural Zimbabwe in the 1990s, it was unheard of to call farmers to a central point for purposes of conducting a household survey. However, it had never crossed my mind that for female researchers, doing things 'the right way' could put them in vulnerable and compromising situations. This encounter taught me a cardinal lesson and for the rest of my fieldwork in 2007, I made sure that someone accompanied me whenever I went to interview male farmers at their homesteads.

Fast forward to 2019, I was meeting with a PhD student I supervised. She was reporting back on a household survey that she had completed. In a casual manner, she mentioned that she had interviewed farmers at a central point using the questionnaire. I asked her why she had not interviewed the farmers at their residences. She was emphatic and highlighted that she did not feel safe at all. Whilst I was initially tempted to lecture her on the importance of interviewing farmers at their homesteads, and that it was a sign that we respected and valued the farmers' time, I remembered my own ordeal. Further discussion on the student's security concerns revealed that indeed that was an emerging major issue for researchers in development practice. My student went on to explain that during some of the focus group discussions that she had conducted, there were farmers who carried guns. What I have also found interesting is that an international research colleague told me that some agencies from the west forbid enumerators entering farmers' homes to administer survey questionnaires. They recommended inviting farmers to a central place for the purpose of conducting the interviews. Given these contradictory positions, does this not justify rewriting household survey handbooks, particularly for places where personal security for women might be a major concern?

Conclusion

The stories I have shared here reflect some perception changing experiences when conducting field research and include unexpected events and lifetime lessons. Working within a multi-cultural and international team set up further challenges one to change roles and identities whenever the need arises. One can never predict what to expect when conducting research in the field. This makes it imperative to be ready for surprises and unique lessons. My experiences highlight the need for scientists from developed countries to be cognisant of local norms and protocols when conducting research in rural communities in Africa. The question is, how can the mindsets, attitudes and behaviours of western experts be transformed to live up to this expectation? Equally, the stories reveal that it might be time to carry out a 'surgical' and incisive review of community-based research methods, books and other resources, ensuring that contemporary issues are infused in the study design. During such re-engineering and reconfiguration of research methods, the dominant perspectives of international technical experts should not override local imperatives.

Petronella Chaminuka has been involved in several international collaborative projects since 1995. For the most part, the fieldwork experiences in those projects have provided an impetus for her to assume multiple identities in her fieldwork. She holds a PhD in Environmental and Natural Resource Economics from Wageningen University in the Netherlands and has more than 20 years' international experience working as a lecturer, capacity builder and researcher. She is currently heading Economic Research in the Agricultural Research Council of South Africa. Her research interests are in rural development, social aspects of biodiversity conservation, food security and rural innovation systems.

Chapter 2
Just Imagine Okavango: An Account of a Rural Development Researcher in Botswana

Oluwatoyin Dare Kolawole

Introduction

Reflecting on my early days as a rural development researcher and now as an academic, I realise that these two phases of my career path cannot be divorced from each other. The two periods are seamlessly and intricately connected as one thing led to the other. Armed with a newly acquired master's degree in agricultural extension and rural sociology in late 1996, I commenced my professional career in rural development as a Junior Research Fellow at the Centre for Rural Development (CERUD)[1] based at the Lagos State University (LASU) in Ojo in Lagos, Nigeria. Designed as a research hub for solving rural problems and providing advisory services on rural issues, the centre was eventually relocated as it was later discovered that situating it within a university environment made the clientele feel inferior to university 'eggheads'. That early experience made a whole lot of a difference in my disposition towards rural issues. I earned my doctorate five years after joining CERUD and went ahead to acquire another degree in Development Studies in England. This was a few years after I had resigned from the centre to join the dedicated Faculty of my alma mater in southwestern Nigeria.

My desire to make a difference in rural life stems from my rural background. Although conferred with a considerable advantage over many of my friends and colleagues in the neighbourhood in those early days, my choice and desire to stay close to my agrarian maternal grandparents and live with them may have robbed me of some measure of a 'luxurious lifestyle', which I could have enjoyed had I elected to stay with my father. Nonetheless, living with my grandparents eventually

[1] CERUD is a quasi-academic, research institution housed within an Ivory Tower where its primary mandate was to conduct impactful, rural development research, whose outputs are directly meant for rural consumption within the jurisdiction of Lagos State.

O. D. Kolawole (✉)
Okavango Research Institute, University of Botswana, Private Bag 285, Maun, Botswana
e-mail: TKolawole@ub.ac.bw

T. Madzivhandila et al. (eds.), *Development Practice in Eastern and Southern Africa*,
https://doi.org/10.1007/978-3-030-91131-7_2

fuelled and drove my passion for enhancing rural wellbeing and progress. This early experience significantly influenced my choice of career path even though I had ample opportunities to pursue other disciplines.

Two decades down the line, I reminisce on my journeys since my early days as a quasi-academic at CERUD and later as an academic in my alma mater, and then as a researcher and academic in a southern African university. I have a strong feeling that this medium might provide the right platform through which I share my personal reflections. Throughout my journey in community and rural development, I have discovered that it does not take much effort to make people in rural areas happy. Equally, it takes little effort to make them angry or sad if promises are not fulfilled, and the clientele's norms and values are overlooked or belittled. Thus, reflection provides an account of the experiences, which I acquired over the years in a peculiar socio-ecological environment. Specifically, it highlights the precarious encounters and successes witnessed in the process of implementing interventionist rural development projects in the Okavango Delta of Botswana.

Enter the Okavango Delta

After eight years of rising through the ranks at my alma mater, the desire to learn new things and explore another academic clime influenced my sojourn in southern Africa. Against all impediments, I joined the service of the University of Botswana at the Okavango Research Institute in north-western Botswana as a Senior Research Scholar late August 2010. Among many others, my job specifications clearly indicated that I should actively undertake rural development and other related research in the Okavango Delta. I am also expected to teach and supervise students, offer services to the university, community and my profession. Taking up a job in a hot, semi-arid and dusty environment such as the Okavango Delta is a matter of choice. Raised in a tropical rainforest environment, my determination to make a shift fuelled that desire. Having been a good geography student, I was aware of the challenges associated with living in a hot, arid desert. As a young man who always liked to live in a very cool and clean environment, which I still do, I knew what it meant to head to a windy and dusty environment that is so remote and where the trappings of urbanity are almost non-existent.

My wife was another influential factor in my decision to move to Botswana. Given her penchant for researching things and places, she sarcastically warned and asked me if I was aware that donkeys and goats were always a common feature of the roads in Maun. I took her warnings with equanimity and determination so much so that she supported me to break these new grounds at the time. Eventually, I found myself in Maun, which is situated at the distal area of the Okavango Delta.

The Okavango Delta presents a unique aquatic environment in the middle of the Kalahari Desert. Unlike most deltas, which empty their water into the sea, the Okavango Delta is an inland delta that empties into the Kalahari sands at the distal area of the water body. The delta's pristine nature and richness in biodiversity (both flora and fauna) engendered its recognition as a Ramsar Site of international importance in 1997 (Department of Environmental Affairs: DEA, 2008) and was inscribed as the 1000th UNESCO World Heritage Site in 2014. The delta is an alluvial fan with distributaries covering a land area of about 5300–15,000 km^2 depending on the magnitude of water flow in a flood cycle (Gumbricht et al. 2004; Mendelsohn et al. 2010). The water flow originates from both Cuito and Kubango Rivers, which derive their water source from the upland plains of the tropical rainforest in Angola. The two rivers eventually converge in the north-eastern part of Namibia to form the Okavango River, which criss-crosses the area and then flows through a fault line to enter Mohembo in north-western Botswana in April. The water then flows through different contour lines to form several distributaries, which reach the distal area of the delta around June or July during the annual flood cycle. Indeed, the dynamics of the Okavango Delta are such that it receives water flow during the dry winter season spanning May and August, and then recedes during the rainy season, which occurs from September to April. Given the uniqueness of the Okavango Delta and its diverse wildlife population in addition to other natural resources, many international and national organizations, including regional governments, continue to focus their attention on the sustainability of this aquatic environment because of its huge tourism potential. More importantly, the communities in and around the wetland, which are predominantly rural, depend solely on the aquatic and terrestrial resources for their livelihoods. These communities have diverse ethnic backgrounds and cultures built around the dynamics of the delta. Besides the dynamics of the flooding pattern and scenarios associated with the Okavango Delta itself, the interactions between humans and this aquatic environment present enormous opportunities and challenges, which require effective management and for which scientific research is dedicated. Thus, the Okavango Research Institute (ORI) is established to fulfil these goals. The following subsections present my case by case encounters as I carried out research in the Okavango Delta.

The Research Expeditions that Almost Went Awry

I have always loved to learn new things and visit new places, be they local or international. It is indeed gratifying to discover that social science research could achieve all this for me. I led a research codenamed 'Weatherman' in the Okavango Delta from 2011 to 2012. The study, which focused on indigenous weather forecasting and farming, was funded by the United Stated based National Science Foundation (NSF). This project took me and my research team to the remotest places in the delta at the time. As we moved from one rural community to the other, we enjoyed every bit of our experience and interactions with the riparian community. This community valued the innovative approach we had introduced in studying their weather forecasting skills. They were particularly ecstatic because of how we engaged them in every phase of the project. Through extensive and stepwise consultations, we made sure nobody was left behind in the scheme of things. Besides, we created platforms through which farmers had the opportunity to interact directly with climatologists and weather scientists during the research process.

We embarked on one of the research expeditions at Tsodilo and Chukumuchu communities to conduct a field survey during mid-February of 2012. This was after we had earlier made appropriate consultations with the two communities, sensitizing them on the objectives of our proposed research. As we would normally do, all the research team members would go in a convoy to the study site, start the survey process with our well-trained research assistants and two technical staff in order to ensure that all was in order. After a day or two, we would then leave the research assistants and the field officers to continue with the field survey. Such was the scenario on this pleasant Tuesday morning when one of my colleagues and I were returning to our base at the university in Maun.

While driving an off-road vehicle, we were a few kilometres away from Tsodilo village when suddenly a very huge elephant that appeared to have been ostracized by its mates decided to charge at us. Simply unprovoked, the aggressive bull jumped out of the bush into the road, crossed to the other side and then immediately turned back to attack us straightaway. The consternation expressed by my Motswana colleague prompted my adrenalin, making me summon courage to further stamp on the accelerator. His yelling—'…this beast will kill us'—was a warning signal that the end was not too far away. We escaped by the whiskers and not because I knew how to manoeuvre the vehicle but by sheer luck.

A few weeks down the line, we had a similar experience that almost proved fatal during another expedition. The research team had gone to the Jao community, which is situated at the heart of the delta where accessibility has always been a challenge; we could only access the village using a motorized boat. The other alternative was to fly a helicopter. Against all precautions to avoid travelling on the water during late hours, we decided to return to our base in Guma Camp that day as we had unavoidably finished our activities in the village very late. It began to rain as soon as we set out on our return trip. Given the nature of the terrain, we would have to navigate through narrow channels shrouded in reeds before accessing the big lagoon on the other side. During the navigations, we were ambushed by a crocodile, which almost capsized our boat, in its bid to subsequently feast on us. Miraculously, we once again escaped the wrath of an embittered creature who felt violated in its pristine environment by the sounds of our boat engine. It then dawned on me that we had escaped animal attacks on two occasions without any insurance cover for the job. Imagining what could have happened to loved ones if we had lost our lives remained conjectural. Reminiscing on these two unpleasant incidents now, the occupational hazards that development researchers are daily faced with in an environment infested by very dangerous animals, which are naturally wired to kill, come to mind.

They Only Come Here to Collect Information!

Driven by the desire to enhance farmers' livelihoods, we began to implement an integrated soil fertility management research in the Okavango Delta in 2012. We visited Chanoga community—situated along the distal area of the delta—during one of our preliminary interactions with selected smallholder farmers whom we had identified as partners in the research. Having met a few farmers, we then proceeded to have a chat with another prominent farmer in the village. As soon as we met the man who was in his late 60 s or early 70 s, the first question he asked us was about one of our colleagues who had long left the institute. Although not showing any annoyance, he retorted and was quick to tell us that an expert came to their community promising heaven and earth about solving a farm problem. The man continued: '…As soon as he collected all the information, which he wanted from the village farmers, the fellow never came back again. Researchers only come to our community to collect information and then disappear into thin air'. I could immediately see the disappointment in the man's face and I also was able to discern straightaway that the man was sending a message to us about their lack of trust for researchers whose motive is only to promote their own selfish interests at the expense of their clientele system. There is no denying the fact that this man's viewpoint is commonplace in most study sites when considering the nefarious activities of some development studies researchers and students who collect information from communities and never go back there to provide feedback on the outcome of their research.

While the man was gracious enough to hear us, I could see through him that we needed to deal with credibility issues in our daily walk with community people. Of course, the man eventually did not actively participate in our project even though we tried our best to convince him that our approach was different from those they had experienced in the past. The lesson derived from this encounter is quite significant. As development workers, we need to detach from any perfunctory obligations that serve our clientele no good. Rather than engage as an academic jobber, we would need to go beyond personal incentives and interests by exhibiting empathy and upholding the virtue of probity to make a difference in the lives of community people if only to succeed in the end and be remembered as an agent of 'good' change. We went ahead and succeeded in that project, but that unpleasant memory ever remained with me. The success of this project is shared in the next subsection.

We Really Cannot Quantify What You Have Done for Us

In our bid to improve farmers' lots through our soil research in the Okavango Delta, we used a 6-level approach to implement a soil fertility project that the beneficiaries eventually considered very worthwhile. First, we engaged in a consultative process by sensitizing farming communities about the aim of the research. Second, we conducted a social survey among selected farmers. Third, we organized some consultative workshop series to provide feedback on the results of our field study. Fourth, we partly employed the service of the Soil Analytical Laboratory at the Department of Agricultural Research in Gaborone, Botswana, to carry out a series of soil tests in selected farming' fields with a view to determining context-specific solutions, which were best fit for specific locations. The standard soil analysis report, which was produced by the laboratory, offered some guides as to which fertilizers and soil amendments were appropriate for specific farming' fields or areas. Fifth, we then liaised with farmers and agricultural extension officers who of course started the research endeavor with us from the outset, and sixth, we implemented the intervention. We ensured that the officers were actively involved to enable them to assist farmers in having easy access to advisory services and fertilizer supplies, which were appropriate for their farm needs. This innovative approach worked for farmers who saw our research team members as some messiahs. That unique experience put the farmers on a pedestal, which they had hitherto not seen in that fashion. They now know that there are specific fertilizers, which are most suitable for their fields, and that they cannot use just any fertilizers to achieve their farm goals. And for them to have unanimously remarked '…we really cannot quantify what you have done for us' did not come to us as a surprise.

Just like any development initiatives that have their downsides, we encountered a few challenges in our bid to further the project objectives later in the course of the action research. Not satisfied with the progress made so far, we decided to carry out some in situ experiments at the Rural Training Centre based at Nxaraga village to test the efficacy of organic soil amendments (such as BioWash, Biochar and elephant dung) in improving soil structure and *cation* exchange capacity (CEC). If the experiments succeeded, they could minimise the impacts caused by baboons and other wild animals in the area, which kept uprooting our crop plants in addition to harsh weather conditions. After so many trials, much efforts and resources, we gave up on the field experiments while devising other strategies, which we could use to circumvent the drawback. This unpleasant experience taught me another lesson. Doing development research and striving to solve a specific rural problem may not be as straightforward as one might think. And it would take a measure of fortitude to follow through with it in the end.

The Plot Lies on a Flood Plain

In April 2012, a colleague who is a technical officer at ORI and with whom I had worked with, walked up to me and said the Land Board recently and unjustifiably stripped his father of the possession of a plot which he had owned and farmed for many years presumably on the ground that the plot lay in a flood plain. He expressed his disenchantment with the government officials' high handedness in issues relating to land access and ownership in rural Botswana. After listening carefully to his story, I decided to write an opinion piece about the matter in one of the popular national newspapers—*Botswana Guardian*. In that piece titled: "*The plot lies on a flood plain"*, I aired my opinions on how land matters should be handled by the institutions responsible for land allocation, particularly in relation to resource-poor people, published on the 4th of May, 2012 (see Kolawole, 2012). Without telling my colleague what I had done, he stumbled on the article and read it, and then called my attention to the publication. About two weeks after the publication of the opinion piece, there was a reversal in the decision taken on the confiscated plot, in favour of my colleague's father. The same individual thereafter came to me and in jubilation thanked me saying that the article, which I recently published, worked in their favour. Although certain that the issue came to public awareness, whether my article indeed influenced the Land Board to reverse their decision on the plot is another thing altogether. What mattered was that I was able to use my expertise to communicate land issues that may have significantly contributed to turning around my colleague's situation. This experience is probably a testament on the importance of community service that many of us may have taken for granted in many cases.

Conclusion

Many years have gone by and, I have now attained a full professorship status in the Academy while the experiences garnered over the years still linger on. Not only that. The desire to make a difference continues to drive my motivation to impact positively on rural communities in the Okavango Delta and beyond. While many opportunities are always available for the taking in the frontiers of rural development research, corresponding challenges clearly abound as well. It is much easier to see impediments to progress (most especially in a difficult, geographic terrain) than to see possibilities in following through with the calling. It takes determination, resilience and purposefulness to succeed on the job. The rural communities in which I have worked continue to appreciate the services of experts and the difference that they make in their life. This in itself is enough motivation to propel us to do more.

References

DEA (2008). Okavango delta management plan. Department of Environmental Affairs, Gaborone, Botswana, p. 13. *Online document*: https://www.ramsar.org/sites/default/files/documents/pdf/wurc/wurc_mgtplan_botswana_okavango.pdf (Accessed 8 November 2019).

Gumbricht, T., Wolski, P., Frost, P., & McCarthy, T. S. (2004). Forecasting the spatial extent of the annual flood in the Okavango Delta, Botswana. *Journal of Hydrology, 290*, 178–191.

Kolawole, T. (2012). The plot lies on a floodplain… opinion, *Botswana Guardian*, 4 May, pp. 16.

Mendelsohn, J. M., vanderPost, C., Ramberg, L., Murray-Hudson, M., Wolski, P. & Mosepele, K. (2010). *Okavango Delta: Floods of Life*. Windhoek: RAISON, pp. 9–139.

Oluwatoyin Dare Kolawole is Professor of Rural Development at the Okavango Research Institute (ORI), University of Botswana based in Maun, Botswana. He was a CODESRIA laureate in 1994 and 2006 and is an Adjunct Faculty at the College of Health and Social Sciences in Eastern University, St. Davids, Pennsylvania. He works at the interface of science, policy and agriculture in Africa with major interests in politics of knowledge, sustainable agriculture, natural resources conservation and community development. He was University of Canterbury Visiting Canterbury Fellow in 2014.

Chapter 3
Ubuntu Social Values—Collecting Research Data

Hlekani Muchazotida Kabiti

Introduction

Having worked in the rural development sector for more than 10 years, I have grown to appreciate its vibrancy and complex nature that requires the application of systems thinking. This sphere of work is complex as each rural development experience present unique issues. The need to tailor make programmes/activities based on contextual, current and cultural realities is obvious. This makes rural development practice both exciting and frustrating.

In this contribution, I present an array of my personal lived experiences during data collection projects in Mashonaland East Province of Zimbabwe, and in Limpopo and Eastern Cape Provinces of South Africa. Through the experiences I learnt that thorough planning is crucial for successful field data collection. Note that even with thorough planning things can still go wrong. Knowledge of the physical environment (geographical terrain) where the research will be conducted is crucial when making logistical arrangements. In addition, given the humility and respect that respondents often display during social science studies it is important to respect local norms and the culture. Lastly, researchers should know that an innovation can emanate from some challenges. Thus, one should always be ready to learn.

Where is the Headlight?

In May 2016, I was part of a team conducting project evaluation studies on smallholder dairy farming in a rural area in Mashonaland East Province of Zimbabwe.

H. M. Kabiti (✉)
Risk and Vulnerability Science Centre, Walter Sisulu University, NMD Campus, Mthatha, South Africa
e-mail: hkabiti@wsu.ac.za

T. Madzivhandila et al. (eds.), *Development Practice in Eastern and Southern Africa*,
https://doi.org/10.1007/978-3-030-91131-7_3

Prior to that field research trip, I prepared copies of data collection tools, recruited and trained four research assistants, developed schedules for household surveys and planned for our transportation to and from the data collection site. It was a Monday morning when we kick-started the data collection process. Everything went as planned, resulting in us conducting interviews in six households. It took about one hour to complete each interview. At the end of the first day, we regrouped and reflected on the day's work. We agreed on how to improve our efficiency and compiled the households we would interview the following day. With that elaborate preparation, we were confident that we would complete our data collection that same day if all went as planned.

On the second day we went to interview the first two households. We drove through the rough and bumpy gravel road that connected the village where the households were with the main road. It did not take long to realise that my sister's "sport utility" vehicle which we were using was unsuitable for such a road. When we planned for the field research, we had assumed that any car with a decent ground clearance would be appropriate for the gravel road. How wrong we were! Part of the journey had to be completed on foot. We walked almost two kilometres to reach the households.

After conducting interviews at the households, we retraced our footsteps to the main road and proceed to the next village. Just before getting onto the main road, we stopped to rest briefly. As I browsed through the household map, one of the research assistants noticed that one headlight of the vehicle was missing. As my eyes quickly drifted from the map to the front part of the car where the headlight was supposed to be, I could not bear the feeling of damaging the borrowed vehicle. It was a sickening feeling as I did not want to disappoint my sister. We started driving back to look for the headlight. After driving for about three kilometres, we found it. What a relief! It was clear that our choice of vehicle for the field visit did not match the nature of the road infrastructure and geographic terrain. If we thought we were done with drama, the next experience only waited to deliver yet another free lesson to us.

Car Keys in a Pit Latrine

We arrived at the next village where we split into two groups to make it faster to interview the next two households. One of my research assistants recognised the head of household at one of the households. He was her former teacher at school. Both were so excited to meet again. It was not surprising that the former teacher was so welcoming to us. I proceeded to the next household with another research assistant.

We finished our interview and went to regroup at the former teacher's homestead. The team we had left there was still carrying out its interview. Seeing that it was taking rather long to complete the interview, I decided to take over. It was easy to continue with the interview because the main respondent was the wife who was most involved in dairy farming activities.

As we concluded the interview, I noticed that the two sons at the household were moving up and down restlessly. I did not suspect any problem that would directly affect our work eventually. After completing the household interview, I went on to ask for directions to the next household whilst also moving towards our car. A soft and calm, "please *wait*", caution came from one research assistant. I wondered and wanted to know why we had to wait. She whispered to me that she had dropped the car keys into the pit latrine by accident. At that point it became all clear why the two lads had been moving restlessly across the yard. The elders in the household had not been informed about this mishap, so I gathered. Thus, I went on to break the news to them myself.

The two sons frantically tried to hook the keys out but to no avail. Everyone at the village tried to solve the problem. It was clear that the method which we were using was not going to yield any results. In general, the atmosphere was sanguine. We did not have a set of spare keys, implying that a solution to the problem at hand had to be found.

Some young men from the neighbourhood heard about our dire and desperate situation. They came to witness and assess the potential for them to help and presumably make some money in the process. The former teacher dashed their hopes of making that money out of our desperation. Nevertheless, the young men continued to offer some help.

After what seemed hours of fruitless attempts, it was recommended that the only way to retrieve the keys was to partially demolish the pit latrine and pump out the faecal matter. Having damaged the car also, this more saddening news to us. At first, the household head did not buy the idea of damaging the pit latrine and pay for repairs. He still believed that we could retrieve the keys without having to destroy anything. Soon, our idea carried the day. Although it was hard work that had to be carried out with caution, we executed our plan. I can still feel the relief that swept through me when the keys were back in my hands. We thanked the family profusely before driving off. It was a dark day but with major lessons. We learnt that *in development practice, an innovation can emanate from a mishap. The better idea to open part of the toilet (one metre deep next to the toilet) without destroying the whole structure saved the day.*

Fitting into the Cultural Context

In 2014, I was part of the Institute for Rural Development at the University of Venda in South Africa's Limpopo Province. We were implementing a community engagement project called *Amplifying Community Voices* (ACV). On this day, we were celebrating Heritage Day in a village located about 40 km to the south east of the university. The ACV had a long-standing, working relationships with the community we were to spend the day with. The university representatives and the community leaders had worked together to formulate a programme that fostered learning about the local heritage, tradition and culture. When we got to the venue of our engagement, almost all the local community members were putting on the Venda traditional attire. A few of them wore Tsonga traditional clothes. The university group comprised about 15 students and two academic staff members. Only a few of us from the university wore traditional attire. We had not thought about doing so.

The event kicked off well with traditional songs and a series of demonstrations of some practices. One practice entailed greeting and paying homage to the traditional leaders. Every speaker did this before delivering what was expected of him or her as shown on the programme. Men were expected to squat. Women lied on the ground in line with the Venda cultural tradition. On the day, I had been tasked to explain the purpose of the day's activities.

I was not familiar with the Venda traditional culture. Thus, I had to study carefully how others were paying homage. In my own way, I thought I had mastered it and good to go. When my turn came, I lay down with my hands outstretched, palms together as if praying and placed on top of my head. The ululation that followed in acknowledgement of my greeting made me somewhat proud. However, there were others who burst who into laughter. At that point some doubt enveloped and rather confused me. I pulled myself up and got on my feet. An elderly man stood up to highlight that it was not acceptable in the Venda culture for a woman to pay homage to traditional leaders whilst wearing pants. More laughter from the audience greeted this explanation.

It was gratifying that the elderly man had found it necessary to correct me, which was befitting on this day when we expected to learn about cultural practices. Even though the elderly man was friendly and civil, I still could not help it. I felt like I had been slapped on the face with mud in full glare of that audience. Summoning some courage, I continued to explain why were gathered on that day. In previous community gatherings we had participated in, no one ever emphasised that we dress suitably for events where traditional leaders were present. This incident taught me that in development practice respecting local norms and culture are cardinal when conducting research and related development work.

Identify the Real Leaders and Verification is Imperative

In June 2019, I was part of a rural development project which sought to bring climate change awareness to local communities and schools in the rural areas of the Eastern Cape, in South Africa. The Risk and Vulnerability Science Centre at Walter Sisulu University implemented the project. To enter the local communities, we contacted a community champion through one of our implementation team members. This community champion claimed to be well acquainted with the local realities, including senior secondary schools in the area. We had planned to engage senior secondary school learners. Our community champion told us that he had all things under control and had a solid network with relevant people in the schools and the Department of Basic Education.

As the day for our first engagement with the Port St Johns community approached, we requested the community champion to provide the contact details of the Principal of the School where our event was to be held. He did not deliver. Nevertheless, he reassured us that all was in order because he had secured permission to host the event. Apart from this, he pointed out that other schools from the area had confirmed that they would actively participate. We felt we had no reason not to trust and take him for his word.

The day of the event arrived. We travelled to the venue of the engagement. Only 12 learners from the host school were released to participate in our event. One small classroom was availed for our activities of the day. How was this going to work given that several stakeholders had displays to showcase their work? To make matters worse, our community champion was nowhere to be seen. Nor was he reachable via his cell phone. Where was he? As we fumbled and tumbled on what to do, the head teacher at the host school then recommended that we try to secure the nearby community hall. We took his advice. The hall was bigger and even more convenient than the school premises.

We moved to the new venue hoping that more schools would participate and help us have a more reasonable number of learners. Our hope soared when the education coordinator for the Port St Johns cluster arrived. We learnt that although she had been informed about the event, there was no request to invite the other schools. After discussing the matter with her she committed to serve as the new community champion. Immediately, she contacted the local School Principals and requested that their school participate in the event.

The phone calls unlocked the school and learner participation challenge. In no time the community hall was so packed that we had to figure out how best to accommodate the learners. Six schools participated, which demonstrated that she was the real community champion. She was integral in the network we wanted to work with and had the power to mobilise. Through this event, we learnt that challenges might be difficult to foresee and that reliance on an individual without clout can be risk. It was evident also that had we carried out proper risk analysis and planned how to act we were likely to identify the potential challenge and avoid the embarrassment and shame we experienced.

Conclusion

The fact that "lived experiences" provide a baseline or guideline in decision making, planning and implementation of rural development practice is evident in the stories I have shared. Furthermore, a thorough understanding of the physical, environmental and social spaces, principles and systems that govern an area is required. The essence of social capital, appropriate infrastructure and technologies, critical and innovative thinking, cultural values and leadership in development practice should never be underestimated. Also, it was evident that challenges faced could create opportunities for one to unlearn and relearn.

Hlekani Muchazotida Kabiti is an agricultural economics, rural development practitioner and academic. She has worked on various rural development projects within the contexts of rural socio-economic and community development through institutions of higher education learning (universities). Most of her rural development work was based in South Africa and Zimbabwe.

Chapter 4
Eking Out a Living from the Valley of Death: A Photo Novella of Lake Ngami in Botswana

Barbara Ngwenya

Introduction

I have a hidden passion for photography and used to be fanatical about taking pictures of family and friends during festivities long before the advent of smartphones. In high school, I went mountaineering with friends, gazed at forlorn rural landscapes and rolling hills. These historical narratives captivated my imagination, a snap shop memorialized and conserved through time and space. As a cultural anthropologist doing fieldwork in the spectacular natural scenery of the dynamic ecosystem of the Okavango Delta in north western Botswana, I believe I have been, and still am, living the dream. I have collected and archived a compendium of digital artifacts that span over twenty years.

The more I publish in academic journals, the more my story telling visual eye perishes. There is hope, though. It begins here. The photographic images I share here reflect a progressive compilation of three fieldwork trips I undertook in September and October 2019. Prior to these visits, I was in Sehitwa (Maun Region, Botswana) running a participatory rural appraisal workshop for a Global Environment Facility-funded sustainable land management project in April 2017. During my first visit to the region, livelihood activities around the endorheic Lake Ngami, north of Sehitwa, were in full swing. The Lake derives approximately 80% of its water from the Okavango River feeder system inflows from June–August, and the remainder from local precipitation. This effluent system flowing out of the western side of the Okavango Delta, fills the lake every season. When flood water arrives in Lake Ngami, migrations of fish (especially the eurytopic Tilapia and catfish), into the vacant habitat are immediate. However, during my field visits, the water levels had already started to drop. This photographic narrative captures my experience of the ecosystem services provisioning to human, livestock and wildlife of Lake Ngami.

B. Ngwenya (✉)
Okavango Research Institute, University of Botswana, Maun, Botswana

T. Madzivhandila et al. (eds.), *Development Practice in Eastern and Southern Africa*,
https://doi.org/10.1007/978-3-030-91131-7_4

Concealed Occupational Risks, Affirmative Anticipation

My field trip started on the 5th of September 2019. I went to the flooded ephemeral Lake Ngami in Sehitwa village, (about 100 km from Maun) in the Ngamiland district capital and commercial hub. Traversing this road was no charmer. One of the most memorable times was when the vehicle I was driving overturned and rolled over three times after the Sehitwa area towards Shakawe. I had fallen asleep behind the wheel, mainly due to fatigue. The four undergraduate students I was travelling with got out of the mangled Land Rover vehicle with minor injuries only. The field equipment, pots, plates, gas cookers, plates, chairs, tents, questionnaires, food, books, papers, bags and blankets were strewn far and wide, showing only a small part of the scale of the accident. The students pulled me out of the vehicle, simultaneously earning me the nickname, "oh dearie!" That's all I could say repeatedly as I took stock of what had just happened. My colleague who was driving behind us was lost for words. All he could say was 'ijajajaaa, go rileng bathong?' (Graciousness, what happened?). Nobody answered him. The accident scene spoke volumes on its own. It painted our experience well. What's surprising is that despite my intense interest in photography, I did not take photos of the scene at all. Effectively, I sealed this experience off my mind. This experience illustrates some of the dangers in the development practice field, requiring experts to drive long distances, in mostly bad roads to conduct research.

Thorny Encounters, Dwindling Bounties

In 2010, Lake Ngami was swollen to its glory days and sustained human and wildlife around the area. Top-down, non-evidence-based government interventions-imposed regulations to control fishing through licensing and gillnet size used in perennial freshwater river channels of the Okavango River. Ultimately, in 2016 and 2017 the government banned fishing and export of dried fish to the Democratic Republic of Congo and Zambia. My heart sank when I read the newspaper headlines. It became clear to me that indigenous ecological knowledge of the flood regimes and emergent primary and secondary opportunities to diversify off-farm livelihood sources around Lake Ngami ecosystem services were ignored. The schisms between legally licensed and illegal fishers, local communities and "internal diaspora" fishers coming from various parts of the country, and local and cross-border salted fish traders confirmed this viewpoint. Clearly, whereas motorboats were suitable for net fishing in shallow waters, no thought was put into the use of hand-dug canoes (*mekoro*), let alone salvage fishing using spears due to gradual desiccation of the Lake. Ironically, as we drove past the Lake shoreline (during our first visit in April 2017), I could see clear vegetation signs that the Lake flood extent had receded significantly. Evidence of this was in the form of vast bush encroachment of Mothabakolobe (*Xantium strumarum*), Musu (*Acacia tortilis*) and Mogotho (*Acacia erioloba*). Here I was witnessing the tragedy of the one-size fits-all approach to community development and the dynamics of kin and fictive social alliances. Why is this experience significant in development practice? What major lessons does it hold for policy making, development practice or community-based research?

Almost sixteen months had elapsed after my last visit. The environmental encounters during my second and third visits were bellicose. A strong stench of dead cattle and overfed Malibu stork welcomed us as we drove through a cloud of dust and gloom. The Lake had become a death trap for livestock, crocodiles, horses, donkeys, fish and hippos. The intense heat that I could feel burning my skin under the intense 40 °C El Nino heat wave was, in part, contributing to this.

Within my visits, the transformation of the Lake Ngami ecosystem services was dramatic from a spectacular water body whose surface area was estimated to be 287 km^2 to a mere 3 km^2 or less mud puddle (locally known as *xhobo*) of death and decay. Ironically, tilapia and catfish that migrated into the Lake with inflowing water were proliferating. The productivity was due to the thousands of cattle which roamed the dry lakebed and deposited large quantities of manure. As spear fishers dug out scores of catfish buried in the mud, cattle stood helplessly waiting their turn to die.

From September 2019, the Department of Veterinary Services (DVS) embarked on retrieving dead animals stuck in the mud at the lake. This was meant to prevent potential human and animal health hazards. The joint environmental clean-up campaign involved Malebu stock and DVS. Carcasses were scooped into a mass grave and set alight. By the end of the operation, 400 carcasses had gone up in flames. An average size cow cost about BWP 3000 (USD 300). This meant that local livestock farmers had lost approximately USD 120,000. What a monumental loss and disaster! Yes, livestock carcasses had been retrieved from the mud and burnt in a mass grave. It was so sad to see how thousands of cattle which roamed the dry lakebed leaving behind large quantities of biomass that fuelled productivity of tilapia and catfish now trapped in mud. A tragic scale of economic loss to the livestock farmers and the likelihood of them becoming victims of the poverty trap for generations to come was there for all to witness. Add this to the government regulations that labelled spear fishers, "Illegal" imposters. It precipitated hostile social relations between those in the lake area and anyone regarded as an outsider. This experience highlights the reality and impact of extreme events on the environment and society. This being the case, what then is the balancing act to embrace?

Hostile Social Relations

As the waters in the lake dwindled, the licensed fishers exited the fishing camps giving way to the experienced spear fishers from the villages scattered across the district. To the local Development Trust, these were 'illegal fishers." Environmentalists labelled them, "polluters". Regardless, hand dug wells were now used as a source of water for cleaning muddy catfish. Abandoned wells became death traps for the desperately thirsty cattle. Fishers displayed open hostility to the local farmers, accusing them of neglecting their stock. In the process, they just ignored cattle trapped in the hand dug wells where they died slowly.

Approximately, 100 fishing camps scattered along the lake shoreline existed. Some fishing camps were pitched in the scorching sun while others got tucked under bush shrubs. Spear fishers were openly hostile to any outsiders who they perceived to be government officers. At the same time, the intrusive gaze of researchers and journalists was unwelcome. This meant that probing questions or pictures were not entertained. These announcements were made loud and clear to us as soon as we parked our vehicles. Clearly, the body language revealed that the threat of verbal or physical assault was real. I had to calm the storm when one of our non-Setswana speaking researchers was caught taking pictures. The fishers wanted him to pay. If not, they would seize his camera. Tactful negotiations with the fishermen carried the day. Relations warmed up as researchers blended with fishermen activities through non-obtrusive participant observation such as partaking in salting of the fish, working in MmaTshepo's fish studio and in general, taking advantage of the end of fishing trip excitements. The latter infused lucrative trade and bartering deals among economic interest groups present. This story highlights the value of skill, tact, and innovative strategies when negotiating with communities. In doing so, never compromise the ethical integrity that lies at the heart of such research.

Conclusion

Conducting development-focused research is a hazardous enterprise. Experiences such as what I have shared here are not found in academic writing. Nor are they even found in research methods books. Fieldwork in a wildlife area such as the Okavango Delta is associated with numerous episodic events that never get reported. I have narrated about how I survived motor accidents and dangerous animals such as hippos, elephants, buffalos and crocodiles. As I drove along the receding Lake shoreline, bush encroachment and vast Mogotho woodlands set in motion my frustration and despair. Tragedy of the one-size fits-all approach to and utter disregard for the role of local knowledge emergent community development opportunities was so glaring.

Besides the preceding experiences, flooding of the lake was always good news mainly due to the abundance of fish that improved nutrition and income security for the local communities. My first encounter with the booming fish trade embodied the mantra, a healthy people and a healthy environment. Even more interesting was the observation that although fishing activities tended to be portrayed as male dominated, in the fish camps (regardless of whether they were legal or illegal by the community development Trust), women participated actively. Some were employees whilst others owned boats. There were even self-acclaimed women cross-border dried fish traders. Others found or created a niche for themselves within the fish trade value chain. Creation of an emergency 'photo studio' by Mma Tshepi, who bought fresh and salted fish, highlighted the spontaneity of daily livelihood coping strategies in addition to sales of food to the mud-drenched fishers.

The hydrological pathway of the Lake dictated flood inflow and desiccation, human emotions, feeling of hope and fear, triumph and loss, death and regeneration, competition and negotiated co-existence of despair and tenacity. I observed many ways of how the Lake's historiography reconfigured economic and socio-ecological relations between humans and animals as new opportunities and challenges emerged. Walking through the lake's landscape over time made me get that eerie feeling … I may have to wait for ten or more years for the next flood inflow… and I dread to witness the walk of the last spear fisher standing.

Barbara Ntombi Ngwenya is a seasoned researcher and has conducted numerous field work and research studies in southern Africa focused on sustainable rural livelihoods and development, HIV/AIDS, poverty and food (in) security, community based natural resources management and indigenous knowledge systems and natural resources management. She joined the Okavango Research Institute (ORI), University of Botswana, in 2002, and has been a research scholar for 18 years at the ORI. She has coordinated numerous academic research and graduate/undergraduate training programs and she represents the Institute in various Faculty Boards and government department. Her research profile focuses on human health and environment, and gender equity in access to natural resources and rural development. She has experience in developing and managing international/local trans-disciplinary research programs/projects by providing critical leadership and organizational skills.

Chapter 5
Farming Systems: A Research Practitioner's Viewpoint

Simba Sibanda

Introduction

Having completed my first degree just before the end of colonialism in the late 1970s, my training was largely based on a model of commercial farming systems. Solutions to farming problems could easily be derived from experimental farm-based research, with little need for on-farm research. Somehow, my colleagues and I were not well prepared for the demands that our new government would require from us in relation to how we could meaningfully support smallholder farmers.

Although I had a deep understanding of the smallholder farm situation, having spent at least twelve years of my formative years in the village in rural areas, I could somehow not relate with this professionally. After all, I had spent over four years trying to learn proper agricultural science and now I was being asked to go back to the village where I had tried to get away from by getting a university education.

That said, the government directive was clear and resolute in its agenda, and one had to do the needful. It also made sense that one should give back to the system that contributed so much to what one had become professionally. Government was gracious enough to make opportunities available to train us in farming systems research, which is an approach that was considered the panacea for understanding and serving the smallholder farming system. As part of our training, we had to learn how to do diagnostic surveys of the farming system as a first step in designing research programmes for the smallholder farmers. On one of these training episodes, we visited a village in Murehwa, a communal area, which is about 80–100 km from Harare. We got to the village just after lunch and went on to conduct interviews with key informants such as local extension officers, agro-dealers, a headmaster of a local school as well as focus group discussions with a sample of farming households. Just before the sun disappeared behind the hills, we were ready to head back to the city.

S. Sibanda (✉)
141 Cresswell Road, Private Bag X2087, Silverton, 0127 Pretoria, South Africa
e-mail: ssibanda@fanrpan.org

T. Madzivhandila et al. (eds.), *Development Practice in Eastern and Southern Africa*,
https://doi.org/10.1007/978-3-030-91131-7_5

The village headman pulled our team leader aside and softly told him that we could not leave on an "empty stomach" as this would be contrary to their traditions and values. Hence, we were ushered into this round hut in the centre of the headman's homestead for what I still remember as a sumptuous party. We had visited at the time of the green harvest, implying that all the goodies that one could imagine were laid out for us. We only got home after 8:00 pm that night.

Another memorable incident would happen five years later, when I was part of a team training research and extension officers from the region on how to conduct on-farm research design, starting with diagnostic analysis of the situation in the smallholder farming areas. I led a group of about ten participants to a village that was about 60 km from Harare where we had been holding these training sessions for the past three years.

Our first meeting was with one household, which had been selected by the local extension officer. We passed through a field with an almost ready to eat green maize crop and a kraal with about six head of cattle. When we got to the homestead, we were told that the male head of the household had left on an errand and would not be back for a while. The extension officer had a brief discussion with the wife of the head of the household and we were given the greenlight to go ahead and hold the interview. The interview started by the usual demographic information on the household characteristics, later moving to assets and livestock holdings.

The head of household came back just as we started talking about the cattle holdings. We had to pause while the agricultural extension officer, who was well known to the man, made introductions and explained what we were doing. We could sense a change in the atmosphere right away and the woman stopped talking as the man took over. We then asked him if it was acceptable to continue with the interview, which he agreed to. We decided to go back and reiterate what we had covered so he could put things into context, until we got to the section dealing with cattle ownership. He categorically denied that he owned any cattle, notwithstanding the fact that we had seen the cattle as we came to his homestead. One of the things we had covered in the training was how to politely bring an interview to an end in the event of meeting a hostile respondent. This was one such situation. We moved on to the next item on the list of questions and then politely told the man that we had all the information we needed and then politely bid them farewell.

This was a clear example of gender dynamics at play in the household as well as a changing culture of communities trying to second guess how they could maximise benefits from development agencies. In the discussion, which we later had with the extension officer, it was suggested that the man might have felt that he might be in a better position to receive some aid from our programme if he indicated that he had no cattle. Unfortunately, ours was a training programme with no follow-up development programme, and which was a fact that we had earlier explained to the household so as not to raise any expectations on immediate gains.

Loss of Community Traditions and Values

Twenty years after my visitation to the same communal area, I was again part of a team to assess the food security situation to inform the preparation of an intervention programme for an international development agency. The previous season had not been favourable to agricultural production and there had been widespread crop failure. We had arranged to meet a representative group of 15–20 farmers at a local rural business centre so we could have a focus group discussion about the food security situation. We had conveyed our plans to the local agricultural extension officer who promised that all arrangements would be made based on our request. However, the extension officer, unknown to us, had been taken ill a week before our visit. He, therefore, asked a colleague to make the arrangements on his behalf, leading to a consequent loss of all the details of our request. The community was told that there would be a food security meeting with officers from a well-known development agency, which had previously been providing food aid to the community.

As we arrived at the business centre, the place was fully packed with hundreds of people. My first reaction to two of my colleagues was that there must be a political rally and that if we had known we should have avoided such a day. As we parked our car, an agricultural extension officer walked to us and asked if we were the team from the agency. We answered in the affirmative, after which he said all these people had been waiting for us for the past two hours! Remember we only wanted about 15 people for the focus group discussion. We had to negotiate with the community leadership that this was only a fact-finding mission and all we needed was to get a feel of the situation from a representative group. The people would not budge and wanted to be part of the meeting, for fear that they would miss out on whatever the agency had to offer. After about an hour we finally had our meeting with the small focus group under the watchful eye of the community.

In a matter of twenty years, the community had moved from a self-reliant one that was ready to share what they produced with outsiders, to a situation where the people expected outsiders to meet their needs. The question as to whether we had failed our people as development practitioners then remains.

Action Research On-Farm and Loss of Experimental Designs

During my time as an academic at the University of Zimbabwe, I was involved in pioneering the practice of having students conduct their research work in the village for their masters and doctoral research projects. Everyone was somewhat apprehensive and felt that this was a big risk from our traditional balanced trials that were the norm.

The first experience in this took place in a small commercial farming area called Nharira, which is 150 km from Harare, where the university is located. We had two students working with smallholder dairy farmers conducting action research to improve the performance of smallholder dairy systems. The work entailed a first diagnostic phase where the students went around the farms interviewing farmers to understand the constraints and opportunities in dairy farming, followed by design and implementation of interventions to address the situation. The students had to spend two weeks per month living among the community over a period of about two years. As the project progressed, the students became part of the community and you could sense the close relationships with the farmers every time we had a chance to visit and interact with the farmers.

One of the memorable activities we had were field days that were organised by farmer groups, with farmers presenting the results of their participatory action research. I will never forget the comment from one of the farmers during her presentation. She noted that "these are now our children and they are part of our community. When they first came to the community, we did not tell them all the truth but now we do not hide anything from them". It is noteworthy that the first time they went to the community was to do the initial diagnostic survey. Therefore, if we had based our studies on that initial encounter and not embedded the students in the community, we would have designed interventions on an incomplete story. For effective on-farm research and development, the researcher should be an integral part of the community, not an outsider with no vested interest in the success of the community.

About the same time we worked in Nharira Small Scale Commercial area, we had another student working on a smallholder dairy project in Gokwe South communal area. The project used a similar approach, being an on-farm research meant to improve the economics of smallholder dairy production. The initiative was to experiment with farm-produced dairy diets as opposed to purchased proprietary rations. Half of the households on the dairy scheme were enrolled on the home-based diet, while the other half used their traditional purchased dairy mix. The trial would run for three months and data on milk production and feeding costs would be collected and analysed.

There was a flaw in the design in that all farmers delivered milk to a common milk collection centre at the local business centre. Therefore, the farmers from the two groups exchanged information on their experiences. Secondly, about two months into the trial, we organised a field day to which all farmers and other community members were invited. The preliminary results showed that the home-based diet was 50% cheaper than the purchased ration and resulted in the production of the same amount, if not more milk. In the following few weeks, we lost more than half of the cohort on the purchased ration group as the farmers switched to the home-based diet. This made our statistical analysis at the end of the trial very difficult. The following questions then arise: Was the trial a success. What about the ethics of continuing with the trial where one group of farmers was losing money? Should we have considered compensating one group of farmers at some point? These are difficult questions that on-farm research practitioners would need to consider because they are dealing with real people's lives.

Conclusion

- Over the years, the research and development environment in the smallholder farming sector has changed drastically and people are not willing to do things for "nothing", even if they would stand to benefit from the interventions in the long run.
- Once-off farm diagnostic surveys are limited in their utility as communities may not always provide the required information meant for a proper design of interventions.
- Gender dynamics at the household level in smallholder farming systems are an important consideration, even in diagnostic surveys where one may need to correctly interrogate the system.
- When designing on-farm studies, it is necessary to consider experimental designs to prevent spill-over effects as well as consider ethical issues where some households may forego benefits to enable the studies to take effect.

Simba Sibanda has been a professional agriculturist for over thirty years, having specialised in animal science and later on animal nutrition. His career has taken him to a public research institution, an international research centre, academic institution, development consultancy and lately a regional policy research organisation. He is currently the leader of FANRPAN's Nutrition-Sensitive Agriculture (NSA) programme, which is focusing on how agriculture programmes can deliver positive nutrition outcomes to smallholder farm families and other value chain actors by providing technical assistance to integrate nutrition and implementation of robust, evidence-based nutrition-sensitive interventions.

Part II
In the Front Line of Development Practice

Chapter 6
Tales from the Border

Phathisiwe Ngwenya

Introduction

The greater part of my working experience has been with national, international non-governmental organizations (NGOs), faith-based organisations (FBOs) and United Nation (UN) agencies. During this journey, I have directed and managed programmes and projects in Kenya, Malawi, South Africa, Swaziland, and Zimbabwe. I have been exposed to diverse developmental situations and interacted with a wide range of social and professional groups, including children, youth and adults. I am happy to be sharing some of my experiences as a development practitioner in Southern Africa.

My Pathway Toward a Community Development Practitioner

He stood in front of us, a group of six newly recruited Assistant Training Officers within the then Ministry of the Public Service, Rural Development Training Branch. He spoke for a while about the Ministry and the Provincial Training Centres located throughout the country. My mind quickly flashed back to Esikhoveni Training Centre in eSigodini in Zimbabwe where I had been posted for my first job after leaving university. The excitement was still there after having spent about a school term as a temporary teacher. I had already spent almost two months there, away from the city lights, and its characteristic hustle and bustle.

I remember vividly when he presented the organizational structure within the Training Centres as he paused briefly and switched to body language. Quoting his words, which later became the foundation of my career path as a development practitioner, he said "you are engaged as Assistant Training Officers, you can wake up

P. Ngwenya (✉)
Box AC 1109, Bulawayo, Zimbabwe

T. Madzivhandila et al. (eds.), *Development Practice in Eastern and Southern Africa*,
https://doi.org/10.1007/978-3-030-91131-7_6

tomorrow as the Principal of your Training Centre. It's possible but look at what you would have missed", pointing at the diagram of a ladder that he had drawn on the board. "By missing just one rung on that ladder you would have missed so many jewels in your career path, especially in this field that you are in—rural development. There is the ladder. It's there for a purpose. Take time to climb it. Don't rush in case you miss a rung, and you fall. I can assure you climbing up that ladder within the rural development arena will expose you to a wide range of experiences and you will learn and be a credible professional." These words above are of a Principal Training Officer who, just like us, had started as an Assistant Training Officer but had risen through the ranks to become a Principal Training Officer within the Ministry and based at the Headquarters in Harare (Zimbabwe). This was part of our induction programme. During the six years with the Ministry, and true to his words, I climbed the ladder. In the process I was equipped with more knowledge on rural development theories and amassed a wealth of practical field experiences with grassroots communities. I believe I contributed to rural development taking place in the areas where we worked closely with extension workers, we trained at eSikhoveni Provincial Training Centre. The years at the Training Centre laid a solid foundation for the development work that I did or continued to do later as a development practitioner.

Some key lessons: In rural development, credibility and considerable amount of practical field work or exposure counts. The ladder will always be there in whatever form. Climb it without missing a rung and for sure, you will never have any regrets. It is crucial that you take time to plan your career path. Once you are convinced you want to be a development practitioner, take time to build it. Never forget that for rural development, being grounded in theory and proceeding to have practical exposure in the field should never be separated from one another.

Values and Ethics When Managing Development Projects

"Are you aware that this man sits in one of the UN Committees in Washington DC? Do you want to lose your job?" These were the questions that the National United Nations Agency (UNA) Official whose portfolio included Give a Dam Campaign (GAD) hurled at me in a threatening tone. I was the Coordinator (at provincial level) of this programme.

We had just completed a one and half day workshop focusing on national environmental conservation. The National UNA Office supported our GAD office and staff. Every month, as the Coordinator, I produced a monthly plan that I shared with all members of the consortium, including UNA. In that month's plan, there was a community meeting scheduled to take place in Beitbridge, Zimbabwe. However, my plan did not include this workshop that we had just participated in. It was abruptly scheduled, and we were invited to attend. Obviously, we had honoured it because it was relevant to our work as the GAD Campaign. When we were advised that we were to meet the UN Official and drive together to the workshop at a lodge in the

Matobo Hills, I clearly pointed out that I was supposed to leave soon after the workshop for Beitbridge. This was for a scheduled community meeting. The UNA official responded by saying that "fine we will organise with other members and you will go ahead with your programme".

The workshop ended at noon; thank God this was the perfect time for me to leave for Beitbridge. I communicated my intention to leave. Alas! This triggered a 'war' I had not even suspected would unfold. Uncaringly, the official said, "Can you cancel that meeting? We need the car for use by the UN official". At first, I thought I wasn't hearing him well. Let me spare you the details on the rest of the conversation. What I remember very well is a call from one of the Rural District Council (RDC) Chief Executive Officers saying "when you left, we were told you were going to lose your job. By the way, we offered the UN Officer our car to visit the caves, and Cecil John Rhodes's grave in the Matobo Hills."

I arrived safely in Beitbridge and the following morning the community meeting went ahead as scheduled. After the meeting, one of the members of the District Training Team came to me excited. "This is one of the best and most productive meetings we have had so far in our district". Attendance was excellent because almost the entire leadership of the two wards were here. Community members came in their numbers. He continued noting that "the issues we have battled with for ages were addressed today and dam site was adopted unanimously". He went on as if giving an oral report. I did not say anything but listened intently. I elected not to think about it considering what had happened the previous day with respect to the meeting. Nor did I want to imagine what lay ahead for me.

A year after that meeting, I left the GAD Campaign to pursue further studies. By August 1999, 33 dams and 15 gravity fed irrigation schemes had been constructed. This experience taught me many things. First, I realised that one should never compromise on matters that impart integrity as a development practitioner. It is crucial to have the courage to stand for your own personal convictions in the face of threats from those in senior positions. Lastly, I still believe that one must develop a clear sense of direction not only for one's own work but also for grassroots communities when working with or serving them.

Never Take Communities for Granted!

The officer representing the implementing partner had just spoken. After his speech, there was dead silence for about five minutes. The local Headman eventually broke the silence as he stood up. Whilst pointing his knobkerrie towards the officer, he said "Don't waste our time here. We are not children. Go and tell them we don't want their money. If we are to build this dam here, where we want it as a community, we will do it with our own hands". Immediately after, he walked away and disappeared into the bushes.

We had all gathered by the dam site that the local community and Department of Water Resources Engineers had identified and confirmed. Evidence of all the due process followed was contained in GAD minutes of meetings and reports. On this particular day, the community and the implementing partner were due to sign an operational and funding agreement. However, to everyone's surprise the officer in the forefront of making this happen came up with something totally different. Out of the blue, the international NGO that had agreed to fund the construction of the dam wanted the local community to change the dam site. No valid or tangible reason was forthcoming. All this was against the ethos of the GAD Campaign. The GAD Campaign respected each stakeholder, especially the grassroots communities. In line with this principle, activities were supposed to be carried out in a transparent manner.

After discussions, the local community refused to change the dam site. It was decided that the agreement with the international funder for the dam should be terminated immediately. Instead, the local community would work with the District Training Team and approach a local large-scale commercial farmer to source the machinery required to dig the core trench of the dam. The members of the community would clear the site, dig the trial pits, pitch stones and grass the dam embankment.

The completion of the dam in good time to harvest the first inflow during that year's rainy season remains one of the most significant developments that highlighted that external 'agents of change' should not impose decisions on communities. The sense of ownership of the dam and commitment that the local community displayed is a lesson for others that might experience a similar situation. Strong leaders who do not compromise when not respected are required to facilitate the change process. We learnt from this experience that although money has always been a transactional tool for corrupt practices, in this case the people stood by their ability to self-drive the dam construction. Lastly, building a partnership of equals amongst the non-equals is not easy. It calls for extra dedication, self-denial and selflessness of all partners in order to accomplish the desired outcomes.

How Authentic is That Status Quo Report You Are Reading?

My office phone rang and quickly rushed to pick up the receiver. These are the words that came through the earpiece, "*Qiniso lonke Mrs Ngwenya, ngiyaxolisa ngokwenzekileyo izolo*", which translates to Mrs Ngwenya, I would like to sincerely apologise about what happened yesterday". This was the District Agricultural Extension Officer for Umzingwane District.

During this period, I was the Training Officer in charge of the Project Planning and Management course that was underway. At Rural Development Provincial Training Centres, we trained Government Extension workers and other Development workers from NGOs and other agencies. We covered four core courses, namely *Rural Development; Project Planning & Management Theory and Practice; Training of Trainers and Communication for Rural Development.* Training in these courses included both theory and practical work, where applicable. For Project Planning & Management and Rural Development we organised field visits, learned and applied some of the theories that were taught and discussed in class.

About a month before our planned field visit, I phoned the District Agricultural Extension Officer and asked him to identify two community projects for us to visit. There were 18 trainees in the course. He had assured me that he already had in mind one nutrition garden and a poultry project as possible ones. He made this decision based on monthly reports he had received. These projects were located far apart. In the same telephone conversation, he further indicated that he would get in touch with the responsible extension worker to alert him so that he makes the necessary arrangements for the visit. Moreover, the District Agricultural Extension Officer (DAEO) promised to join us for the visit. I followed up this conversation with a formal letter.

On the day of the visit we got onto the bus with the trainees and three other Training Officers. However, the Vice Principal who had indicated that he would join us could not make it because he had to attend a meeting at the Head Office. On our way, I asked the DAEO about the extension officer in charge of the projects. "Unfortunately, he is on sick leave, but he assured me that all is set, and the project members are waiting for us", the DAEO explained. To cement this, the DAEO briefed us about the projects. We got to the first site, which was the poultry project. The participants were happy to receive us and responded to the questions we had. We made some observations that we planned to discuss in the classroom back at the Training Centre. After that we would give feedback to the DAEO, who would in turn share with the extension worker. We expected the latter to then share our observations with project members. One of our observations was that some statistics that we were given during the briefing session were different from what we observed on the ground.

We drove to the second site, which had been said to be a flourishing nutrition garden. There was no sign of a garden at all. There was a homestead nearby. I disembarked the bus with the DAEO and went to find out in case we missed the exact project site. The mother of the household was indeed a community mobilizer. She knew about projects within the area and the extension worker as well. When we asked about the garden, she told us it did not exist. However, she mentioned that sometime back there was talk about starting a vegetable garden. This was to address malnutrition for children under five. The idea never took off. The garden was only flourishing on paper.

This experience raises numerous issues. It is clear here that the agricultural extension worker "cooked figures" in order to please the reader of the report. Furthermore, if you plan a field visit with external people (donors, course participants, and general visitors) ensure that what is contained in reports is verified. After this embarrassment, what action do you think the DAEO took?

Conclusion

In my work I have come to learn and accept that Rural/Community Development is a complex and multifaceted process that needs to be approached with a lot of tact.

I have learnt that it's possible that once communities detect maltreatment, oppression, imposition/prescription and lack of respect, they can easily mobilise themselves against any type of change. Supporting communities to help themselves is one of the recipes for success in rural development and so is partnership building. It takes partnerships of different stakeholders to bring about true and meaningful community development. Development facilitators should have community passion, dedication, good stewardship of resources, integrity *(have sound work ethics, be honest, dependable, with strong values & morals, be principled, doing the right thing no matter who is watching)* and should support implementation processes as per the agreed upon plans and should not lose sight of their deliberate career progression.

Phathisiwe Ngwenya is a seasoned, result-oriented international development practitioner with more than 30 years of practical work experience in the community development field. She has worked for Save the Children International, World Vision, UNICEF, UNDP, African Medical & Research Foundation (AMREF), Catholic Relief Services (CRS), Adventist Development & Relief Agency (ADRA), national local NGOs and relevant development focused government ministries in different countries.

Chapter 7
Life as a Humanitarian Practitioner

Mafuta Wonder

Introduction

I got married at the age of 28 and by then I was already a development worker. Development work comes with challenges which put tremendous pressure on families. Not being available for the family when you are needed may be disastrous to a relationship especially for young families. For instance, in my career, I have worked on missions where I only had 1 week every 3 months to rest and recuperate. Below I share some of my work experiences as a humanitarian worker.

Language and Culture Barriers

As with any sector, communication plays a key role for effective implementation of development projects. Africa is a continent with a rich and diverse language portfolio. This has both advantages and disadvantages in the context of development projects. On the one hand, development projects bring together experts from countries with different languages and cultures, and indigenous language proficiency can be a major barrier for foreign experts. On the other hand, some local communities might be more comfortable and well vested in communicating in their local languages. The latter highlights how issues of integrity and sovereignty must be considered as people in their own country have the right to express themselves the way they want. Despite the advantage of communicating in English or French, this may not appeal to some communities and is viewed as a western world dominance by some voices.

Whilst working in Somalia, in 2015, I was tasked to represent a developmental organisation that I worked for at a meeting to discuss a directive that had been issued by one of the Ministries. The directive was read in English and stakeholders

M. Wonder (✉)
The University of Venda, Thohoyandou, South Africa

T. Madzivhandila et al. (eds.), *Development Practice in Eastern and Southern Africa*,
https://doi.org/10.1007/978-3-030-91131-7_7

were given time to ask questions and seek clarifications. After a few questions were posed, the meeting became tense and I could hear side meetings being conducted in the vernacular language. The side discussions took more than 30 min and the invited partners were no longer sure if they were to leave the meeting or stay put until the meeting officially ended. Considering that I was not familiar with the different languages in Somalia, I was completely unable to minute the outcomes of the meeting that were required at the office. Frustrated that I couldn't get details of the meeting, I approached one of the government representatives and engaged him in my home language (From Zimbabwe), for about a minute. The official stared in disbelief and had no clue of my mother tongue, but at least I made a point and instantly left the meeting. Two hours later, the government representative called me on the phone and gave me an update in English.

During my work missions in sub-Saharan Africa, I always felt at home as my skin colour and complexion was like any other person in the street. I could go to the market and buy vegetables and fruits with no one looking at me with scrutiny. With the ability to master some language basics, I would freely move around town, get into shops and go on outings during weekends. I would realise work life balance to the fullest. However, my work took me far across the oceans in Asia; Cambodia and Philippines. In these regions, I spent days without seeing a fellow black man. Each time I walked around in town everyone would stare at me. In Cambodia for instance I had an awkward situation when some kids screamed when I tried to lift them up after looking at me. This was because they were not used to Africans in that part of the world. In the streets the teenagers would want to hold my hand just to feel if the texture of my skin. These culture shocks are common when working as a development practitioner, and one has to learn to understand and have a perspective of the local communities, especially if the intent of these gestures is not meant to course harm.

Road Infrastructure and Rural Development

After completing my first degree, I was appointed as a water and sanitation field officer in Chivi, Zimbabwe. Each day, I travelled more than 100 kms to and from Zvishavane to Chivi, supporting the rural communities with dam projects, borehole rehabilitation and the irrigation of group gardens. I would use a motorbike for all my errands. It was mostly on special occasions, like delivery of materials and field days that I would drive a truck to the field. When using a motorcycle, I would use makeshift village tracks or foot paths to move from one project site to the next. The communal tracks were shorter, convenient and would allow me to pass through project beneficiary homesteads. Each of the project members would encourage me to park in their homestead before proceeding, and this way, I built rapport with most households in the project area.

On one occasion, I was informed that one of our donors was going to visit the projects in the coming days. I worked tirelessly to inform the communities and the authorities at Chivi government offices about the coming visit. Using my motorcycle, I mobilised stakeholders across many sectors. A day before the visit there was a heavy downpour and one of the makeshift bridges that links the main road to the project dam site was swept away. Given the time, the community did not get time to inform me about the bridge that had been swept away. On the day of the visit, I led a convoy of cars with development partners from Zvishavane to Chivi. We drove through the road that I had used during my normal working days. Upon reaching the bridge, I was shocked to see that the bridge had been swept away and we still had 3.5 km to go.

Despite the organisation policy that did not allow us to carry unauthorised people in organisation vehicles, I decided to pick one project beneficiary to help us with directions using an alternative route. As much as organisational policies are important, it is vital to note that development workers do not operate in a vacuum, as such leverage should be made between organisational policy and relationship building with the communities we work with.

Using the alternative route took an addition 2 hours for a 3 km journey. The delayed start of the meeting due to the detour resulted in half the community leaving the site as there was an ongoing food distribution exercise that was being conducted by another agency. This story highlights the importance of good road infrastructure and communication in rural areas.

A Case of the Ebola Crisis Stigma

The essence of humanitarian response is to save lives for the affected and to alleviate the suffering in times of disasters. The Ebola crisis in Sierra Leone began when I was already in Kenema district, which was the hotbed of the crisis, working for an international non-governmetal organization (NGO). The district is located to the East of Sierra Leone, close to the border with Liberia. The border crossing is very porous and as such, it was quite easy for infected people to move between the two countries. As there was little understanding of what Ebola was and the causes thereof, a lot of conspiracy theories were raised especially in the first 3 months of the outbreak. Despite the campaigns and advocacy calls, like any other behaviour change programme, it took long for the people of Sierra Leone, to evolve a positive mind-set on how to manage the Ebola crisis. Everything linked to Ebola was handled with a pinch of salt and the stigma associated with Ebola was extremely severe.

Case 1—It was widely reported that hundreds of health workers were affected by the deadly disease in Sierra Leone. In towns like Kenema, the community had confidence seeking treatment at health workers homes rather than going to the hospital.

When authorities were tracing Ebola patients, the families did not comply and hid the patients in compounds. This resulted in high risk of infection for family members who provided care for the patients and subsequently spread to other village members. One's sickness during this time was dealt with surreptitiously thus depriving the clinical and psychosocial support that any patient requires. The stealth attention given to patients by family members shorn of medical background undermined the right to proper health which could also be coined a form of stigma. As a result of such stigma, entire families succumbed to Ebola.

The stigma associated with Ebola resulted in unprotected transportation of patients to hospitals. For instance, during one of the meetings, it was noted that families with Ebola patients paid motorbike drivers, who would charge high amounts due to the risks, to illegally transport patients. These operations contributed to the further spread of the diseases as the motorbikes also ended up being contaminated. When news spread, many were advised to avoid motorcycles in order to reduce the risk that came with contact with the sweat of potentially infected bike riders.

Case 2—After a few months in Sierra Leone, I travelled back to Zimbabwe for leave. As part of the due diligence, to mitigate the spread of transmission, the security authorities thoroughly screened travellers from Ebola affected regions. I was screened at the Lungi international airport in Sierra Leone and Nairobi as I had used Kenya Airways. On arrival in Zimbabwe, the Ministry of Health was also conducting screening. The health authorities were asking passengers if they had been or passed through a country with Ebola in the last three weeks. When I mentioned that I was coming from Sierra Leone, the official froze and dropped my passport in shock. The immigration official who was supposed to stamp my passport was literally shivering when he saw that the last stamp was from Sierra Leone immigration officials. My profile including contact details were captured during a short interview to ascertain my health status. After about 30 min, I was released and instructed to report to the local hospital daily for the first week. It was one of the unique holidays that no one visited me or my family as people dreaded I could be infected. Whilst I had the privilege of just being with my family, inwardly I felt I was a victim of stigma. After a week I decided to drive to Mozambique for a holiday with my family. At the Mozambican border it took me 6 h to be cleared to proceed with my journey into Mozambique. I got accompaniment from authorities at the border post to Chimoio hospital where I had to be screened by health authorities. The first two days during this holiday, I had to report to the authorities for check-ups. As a humanitarian worker, one must reflect on the risks associated with some of the humanitarian response. Some of the questions I asked myself included. Is saving lives synonymous with endangering my own and family?

Conclusion

In the execution of their duties, humanitarian and emergency workers face a lot of challenges, some of which border on stigmatisation. Whilst the workers have an important job to save other people's lives, it is often a thankless job as their efforts are not appreciated but sometimes, they are ridiculed. While executing their assignments, the workers actually expose themselves and their families to very high and real risks of contamination. What the beneficiary communities see is the mistakes and errors that the workers make without understanding the stress and pressure they are constantly under.

Mafuta Wonder is a seasoned environmental and behavioural Scientist. He has 14 years of experience working on development and humanitarian work in third world countries, having worked in Zimbabwe, Cambodia, Philippines, Sierra Leone, South Sudan and Somalia. As part of humanitarian response, he was involved in the deadly Ebola crises, typhoon Haiyan and the drought in Somalia.

Chapter 8
Supporting Small Scale Farmers in Zambia

Idowu Kolawole Odubote

Introduction

I am an Animal Scientist with experiences spanning two sub regions, West Africa (Nigeria) and Southern Africa (Zambia). My involvement in the livestock industry was mostly limited to training of students as future work force, conduct and supervision of research mostly in the university research farm facilities. Interactions were thus limited to the academic circle – students, fellow researchers, conferences and research farms.

All the above changed when I became the de facto team leader of a multi sectoral donor funded research project on scaling-up climate smart agricultural solutions for cereals and livestock farmers in Zambia which targeted 50,000 smallholder farmers. At almost the same time, I was appointed as a short-term expert on another donor funded dairy milk value chain project in southern province of Zambia targeting smallholder dairy farmers. The two projects provided me with the opportunities to travel extensively to the rural areas and these brought me into close contacts with smallholder farmers. The exposure made me appreciate the farmers' efforts in the food production chain despite the numerous challenges that they face daily. The projects also exposed me to the practical elements of rural development using livestock as one of the drivers of eradicating poverty and hunger.

Information is not Free

During one of the field visits to meet farmers, we arrived at the venue and were warmly received. We were about to commence the focus group discussions when the bomb shell question was dropped as a matter of fact: "What do you have for us?"

I. K. Odubote (✉)
Zambia Academy of Sciences, Lusaka, Zambia

T. Madzivhandila et al. (eds.), *Development Practice in Eastern and Southern Africa*,
https://doi.org/10.1007/978-3-030-91131-7_8

That question opened the door for many more questions and comments: "What do we stand to benefit? We know you have been paid to do the work, what is our share? You just want to obtain our knowledge for free without paying for it. Don't be stingy, share?".

We were perplexed. All our entreaties that the information gathering is part of giving back to them to make their lives better fell on their 'open' ears without any impact. It is, however, noteworthy that the meeting was a non-starter.

Looking back now, I believe our project planning and initial contacts were not broad and deep enough for the community to appreciate the goals of the project and nip in the bud any potential mistrust. The buy-in of the project by a wide spectrum of the community should have been sought as the benefits would have been obvious from the onset. Perhaps, it could have helped if we had jumped on existing well accepted government programmes or even the camp extension staff who lived among them.

Tricked into Offering Free Ride

We were in Namwala, Southern Province meeting with the listed smallholder dairy farmers. However, one of the farmers with a fairly large dairy cattle herd was not part of the meeting. From the information gathered, he was key to our assessment as he could provide us with a large amount of information. We were therefore desperate to meet with him and inspect the herd. At the same time, we were conscious of running behind time. We did a rough calculation, that for us to travel back to Choma town (which is a distance of 180 km) before it is dark, we must start off at 4 pm at the latest. So, we reasoned that if we spent an hour meeting the farmer and collecting the necessary information from his herd, we should be fine as the day's tasks would have been completed and within time too.

Several phone calls were made to the last respondent and we kept receiving the now familiar response, which was 'wait for me, I will soon be there'. Since the farmer was not forthcoming, we decided to 'kill time' and called another farmer to make up for the number of farmers visited. The farmer responded in the affirmative but asked us to meet his caretaker who would take us to the farm. Within a few minutes we were with the caretaker and on the way to the farm with him as the guide. Time check was now 3 pm and we were happy with the decision taken which we believed had yielded good results.

After about 30 min driving on the all-weather road, we got concerned. All we could see was bush, but he kept telling us that it was not far! (I have since realized that farmers do not measure distance and time with a standard measurement instrument, it's all about getting to the destination). We continued for another 20 min and arrived at a fair looking brick-built house. Then the caretaker did the unthinkable. He got down from the vehicle and we could tell that he was asking for directions from the fairly old woman standing next to the road. By this time, we had lost about an hour and we still have the 2 hrs drive back to town. We called the farmer but by then the mobile network was no longer reachable. We had a quick meeting and resolved to abandon the visit and head back to Choma. We informed the caretaker when he got back to the car of our decision. Surprisingly, he was not perturbed. He wished us a safe trip and bounced ahead of us singing even before we could make the U-turn.

Were we tricked into offering a free ride? Was the farmer an accomplice? We will never know because when we finally got through to the farmer, he mentioned that the caretaker was recently employed and not so familiar with the directions to the farm. The farmer could be right as the African saying, all roads led to the market; this is equally true of many smallholder farms.

Researcher Turned Business Advisor

We were conducting a genetic survey on the smallholder dairy milk production in one of the districts in the southern province of Zambia. We visited one of the successful lead farmers who, by all standards, have everything many others will envy. The farmer qualified to be tagged as an emergent dairy farmer. He had over 100 herds of cattle with 61 breeding cows comprising mostly Friesian crosses. He had a history of using Artificial Insemination. The body condition of the cows was moderate. He had a milking parlor, a tractor, operated a spray race to combat ticks in his herd and that of neighbours, and a biogas in operation. He was a farmer every livestock extension officer would be glad to have as a client. However, not everything that glitters is gold.

On this day, we noted that the usual hustle and bustle of farm activities were not the case. The environment was quiet and dull. The big man came in with defeat written all over his face. What could be the problem? The lamentations then started. What did I do wrong? Where did I go wrong? Was I tricked into dairy cattle production? It was evident that the man was in serious trouble. He had invested heavily in the dairy cattle production and milk supply business.

Things were rosy and business was booming until recently. His grouse was with the low milk price and the erratic payment from the milk processing companies among others contributed to his worries. The drought of the year had also made feeding his animals an uphill task. Prices of inputs (e.g. feed supplies) had been skyrocketing despite the milk price being stagnant. He mentioned that he last made profit in 2012–2014. It was a cul-de-sac situation. His mind was made up to reduce the herd size. In fact, he had already commenced switching to aquaculture with the construction of four fishponds.

The realities were all there for everybody to see. We tried to encourage him by advising on strategic feeding of his herd and that the fishponds could be integrated with the dairy cattle production. Here is a hardworking farmer caught up in an unfavourable policy environment and policies that are outside his control or influence. We were hoping to showcase him, but he was throwing in the towel. What would be the impact of his decision on the numerous farmers that looked up to him? What would happen to his investments and livelihoods?

This issue was picked up and discussed extensively at the subsequent project review meeting. It was agreed that a business development manager be seconded on the team to evaluate his business model—such as the income streams, areas that require cost cutting, new opportunities to create more income channels. More importantly it was agreed that the cooperative to which he belonged, and the project should engage the Dairy Association of Zambia, milk processing factories; and lobby for higher prices and payment of premium price for butter fat content for milk supplied by farmers. In addition, and as part of a sustainability strategy, the need to engage the government to improve on the business environment by putting up policies to discourage importation and dumping of milk and milk products which depress prices of milk and milk products locally was highlighted. It is good to note that not long after, the milk prices were increased and the premium price for butter fat content was established.

The Young Walking Institutional Memory

As part of the genetic survey of, smallholder dairy milk production system in Zambia, we visited a farmer. As part of the assignment, we were trying to digitalize the calves' data collection hence the need to physically inspect each animal and document the pedigree (i.e., a list of the parents and other relations of an animal) in addition to the morphometric measurements. We did not have much issue with the morphometric measurements as the animals were physically restrained after so much struggle and chasing around the kraal. Our challenge was with the pedigree of the animals. The farmer did not keep records and as such was not so sure about the sire and dam of the calves. He kept mixing the parentage and dates of birth and then muttered some words under his breath which we later got to decode to mean, 'how I wish this boy was here'! Not long afterwards, the walking institutional memory walked in. A boy in Grade 5 (primary school) and about 10 years of age. The mood of the farmer suddenly changed as he welcomed the boy.

The boy was equally in a happy mood, comfortable and at ease as he walked into the kraal still in his school uniform. He took us through each calf's parentage effortlessly and without any struggle and more importantly without any record book. Meanwhile, the father kept smiling and grinning from ear to ear and informed as many as listened to him that the herd belonged to the boy. The boy no doubt took ownership of the herd. He knew the animals like the back of his palm. He understood them and the animals were very much at home (relaxed) with him too. He even rode on one of the bull calves in our presence.

I saw an excellent stockman in him if he continued the same trajectory. The career advisor in me took over and asked the familiar question, what career will you like to pursue in life? The answer was a bombshell. Medical doctor, he responded without batting an eyelid. I felt deflated. When prodded further on why he wanted to be a medical doctor, he shrugged his shoulders and glanced at the father. Here is an excellent stockman without any formal training except acquired skills and interest. Have we lost a good stockman? What prospects does the future hold for him as a medical doctor, stockman or even something radically far off? Could he be a good doctor and still retain the stockman ship? Would he be better off studying animal production?

Further discussion with the father revealed that the boy is the future of the farm and herd being the only child. We became aware that the father sowed the seed of a medical doctor in the boy as a status ranking. When prodded on what would happen to the farm/herd after the boy must have qualified and have started practicing medicine, he responded, he could always come to see the animals on weekends! The father was advised to make sure that the boy was seen by the school's guidance and counselling unit before completing his primary school education to make sure that the boy's future is not compromised or jeopardized.

Conclusion

The stories presented here bring out a few important lessons for the development worker: (i) clarify your expectations and that of the community clearly from the onset, (ii) frictions are better avoided if the project was jointly identified and designed as community would have ownership, (iii) be prepared as much as possible for any eventuality in the field to avoid being caught off guard, (iv) take time to study and be accustomed to the culture of the community. Being able to speak the local language helps, (v) be ready to put on a professional hat different from area of specialization in the course of interaction with the rural community such as a career advisor or business consultant, and (vi) be an encourager even in the face of hopelessness.

Idowu Kolawole Odubote is an animal scientist (animal breeding and genetics expert) by training and lecturer/researcher by profession practice. For more than 30 years he has been in the academic space teaching and conducting research in livestock production—chickens, rabbits, goat and cattle—beef and dairy. His experience spanned two sub regions, West Africa (Nigeria) and Southern Africa (Zambia). He joined the School of Agricultural Sciences of the Zambian Open University as a Senior Lecturer in January 2016 and was appointed Dean of the School and member of Senate in July 2016. He led a CTA funded research on Climate-Smart Agriculture and is currently an Independent Agricultural Development Consultant and Short-Term Expert on GIZ funded dairy milk value chain project in Zambia. He is also consulting for the Ministry of Fisheries and Livestock under the African development bank funded on the Climate Resilient Livestock Management project in Zambia.

Chapter 9
Community-and Self-Empowerment Through Experiential Learning

Alois Sibaningi Baleni

Introduction

In 2000, I secured my first professional job as a Training Officer for the Communal Areas Management Programme for Indigenous Resources (CAMPFIRE). That is when my passion for rural development was ignited. From that time, many changes happened in both my personal and professional lives. I got married and started a family shortly after that first professional job. My desire for professional growth was clear and I was determined to develop further. To live up to this, I enrolled for a diploma course in management. Many professional qualifications were to follow thereafter. I am currently awaiting examination for my Doctoral studies. As my experiential stories reveal, I have endured and enjoyed my development practice in local community, national and regional development in Southern Africa.

Staring at Death on the Call of Duty

Prior to joining the Rural District Council (RDC) in Zimbabwe, I taught Agriculture at three local high schools in the Plumtree area of Matabeleland South. Migrating to Plumtree town to take up a professional job with the RDC was one of the fascinating experiences in my early career. I had access to a vehicle that enabled me to conduct field work. However, the standing policy was that for anyone to drive the RDC vehicle, council principals should certify them. This was an additional requirement beyond being in possession of a valid driver's license, which I did not have at that time.

A. S. Baleni (✉)
Society, Work and Politics Institute/ African Center for Migration and Society, Wits University, Johannesburg, South Africa

T. Madzivhandila et al. (eds.), *Development Practice in Eastern and Southern Africa*,
https://doi.org/10.1007/978-3-030-91131-7_9

One morning, after going through all the procedural council authorisation process for the field visit, I requested a colleague from our Finance Department to drive me. We used a Toyota Hilux double cab specifically procured for the project in which I was a Training Officer. During those years, there was so much prestige attached to driving such a vehicle to the rural areas. Visiting a community project site was another appealing thing as well. For me, the field trips constituted an opportunity to keep in-touch with my old network, and to maintain close contacts with friends from the rural schools that I attended.

The trip was set and authorised. My line manager requested that we drop his family members at his rural home as we proceeded to the project site. This arrangement violated RDC policy which stipulated that non-staff members should not be accommodated in vehicles we used for professional duties. We set off and, on the way, I decided that I should touch base with my best friend at one of the local high schools where I had taught before. All was supposed to go well because the bigger mission was my project work and of course to also deliver my Line Manager's family members. As we drove back to the main road, we saw a small boy seemingly walking away from school. I almost instructed my colleague to pick him up. However, an inner voice countered my thoughts. A few minutes after passing the young boy and approaching where we would have probably dropped him off there was a sharp curve, which my colleague driver failed to negotiate well. Our vehicle flew and completed a 360 degree turn mid-air and landed on its side. All this happened extremely fast. The entire windscreen was thrown away. Given that all doors were locked, the only escape route was through the damaged windscreen. Our driver was in panic mode. He ran all over, mumbling inaudibly. In the midst of the confusion and commotion, my thoughts ran to the fate of our unauthorized passengers. I requested our unharmed extra passengers to leave the scene and find alternative transport to complete their journey. My mind still reflects on what could have happened had we offered a lift to the small schoolboy and have had him at the back of the vehicle which had no canopy.

Developing Skills Through Note Taking and Writing Reports

One thing I realised in the space of development work is the demand for report writing and meeting deadlines. All the projects I was part of demanded adherence to tight timelines when submitting progress reports. That reinforced my skills.

I joined the Bulilimamangwe District Rural Council, in southwestern Zimbabwe, as a junior manager. As such I attended all management meetings. I was fresh from college and one of the youngest members of the RDC management team. We held our meetings every Monday. In my first meeting, the Chief Executive Officer (CEO) asked me to capture the proceedings of the meeting and write minutes thereafter. It was resolved that a rotational arrangement of generating minutes of subsequent meetings would be applied. Yet, I was asked to write minutes of subsequent meetings. Inwardly, I was filled with resentment and felt abused, presumably because I was one of the youngest. It never crossed my mind that this was skills development or on the job training at play.

Beyond attending management meetings, I developed various types of reports, which peers and senior managers acknowledged to be of exceptionally high quality. I was fortunate to have joined the RDC at a time when so much development work was ongoing. Many NGOs were introducing various initiatives in the District. During multi-sectorial or inter-organisational meetings and workshops, I found myself at the centre of writing reports as an official of the RDC. My line manager advised me many times that I must take documentation seriously—even as a career. It never occurred to me, once again like in the case of the management meetings, that this was an opportunity to fully develop my writing skills.

Fast forward, to August 2009, while in rural KwaZulu Natal Province of South Africa on a field assignment. A colleague who was an Office Assistant called me and asked: "where are you? We are being retrenched?" At that time, I was working as an Operations Manager for a Consultancy firm in the field of social development and was based in Johannesburg. Upon returning to the office, a retrenchment letter greeted me. It carried the message that I was going to be retrenched with effect from 1 September 2009. Indeed, come September, I was retrenched. However, not so long thereafter I came across a copy of the *North Eastern Tribune* newspaper. It contained a job advertisement for a Committee Coordinator with excellent writing skills. I sent my resume immediately. An interview followed. I was offered a contract for three months, assuming duty on 1 October 2009. Later, whilst I was still serving my contract the position was advertised as a permanent job. I applied and got the job. No doubt, this was one of the fruits of investment in passionate writing of reports as I worked in Bulilimamangwe RDC. The fact that I was also hired on many occasions to write minutes and reports for various entities of a university where I was pursuing postgraduate studies confirms my argument. All thanks to my superiors who had the ability to spot my abilities.

Leading from Behind: Communities in the Lead

I had the privilege of sitting back and seeing the Integrated Rural Development Programme (IRDP) initiative coming to fruition. It was introduced in Bulilimamangwe District, courtesy of funding from the WK Kellogg Foundation (WKKF) and coordinated by the University of Pretoria. The central WKKF philosophy was "helping people help themselves". After a rigorous initial 3-year phase of tailor-made needs-driven interventions in selected target communities, the time to let grassroots communities lead their own development had arrived. The selected participating rural communities, which were mainly the poorest of the then 35 wards in the district, jointly formed a Community Trust called *Tjinyunyi Babili Trust* (TBT) to serve as the vehicle through which development initiatives would be implemented.

I served as the founding Chief Executive Officer (CEO) of TBT, whilst also heading its health and welfare department. While heading this department we ran various initiatives. For purposes of this experiential "lived" experience, let me focus on the household hygiene and kitchen management project. This project allowed members of households to design and built cupboards for kitchen walls as well as rakes for drying and storing cups and pots. In addition to designing washing sinks, the project supported households to dig/develop refuse pits through creative ways.

The integrated approach to development drawing from the initial IRDP process helped the Bulilimamangwe IRDP site to attract funders. For instance, the WKKF invited the IRDP, to an Orphans and Vulnerable Children (OVC) Colloquium in Cape Town in 2005 to showcase their OVC model. This opportunity was exploring how OVC was integrated and mainstreamed into community development practices. As an Officer in the field, working with communities in that space, I had the privilege to lead a delegation of community members to the colloquium.

The structural design of this OVC model was that at a village level, volunteers were trained in and equipped with several skills such as counselling and bereavement support. In some instances, these included Community Health Workers or Peer Educators. Our skills-base in the volunteers for the OVC project was very wide. It included teachers based in the schools in all the beneficiary communities. At an implementation level (i.e., village level), selected volunteers were attached to sizable households clustered together. At every village, there was a granary scheme, where through Village Development Committees and Village Heads, grain was collected from households and other donations for central keeping. At the pilot ward, the advantage was that the Chief of the area was resident in the ward and very much supportive of the OVC initiative.

The uniqueness of the OVC model was in its integrated approach with the traditional clan system for implementation and ownership. Our volunteers were able to visit child headed homes. Amongst others, fruitful results that emerged in the case of pure child headed families, include some of our volunteers moving into being foster parents using the local traditional system overseen by traditional leaders. This model caught the eye of the donor through visits as well as through the reports I compiled. The model was indeed identified as a best practice worthy sharing with the 7 Southern African countries funded by the WKKF then; hence the invitation to the colloquium. Empowered community members prepared the presentation and it was showcased for learning and sharing. The empowered communities oversaw their initiatives. At some point this model community started driving the rural industries economy where a Village Bank system was in place and community members started applying for loans. This is one of the stories that I was privileged to see germinate and grow while I was working with communities in impactful development.

Conclusion

What is important to note about the stories I shared is that knowledge generation and possibilities for change and real self-empowerment through experiential learning is with the communities. Most communities, particularly the rural-based ones are usually side-lined at the expense of elite or urban communities and such development tends to miss the rich texture of knowledge. Rural-based indigenous knowledge systems have the power and advantage of being African and safer from Eurocentric thoughts not that the latter is anything wrong but when seeking local solutions, the former is the best fit. What is seen from these stories is the power of locally based initiatives. Participatory community-based interventions can also be best models of action research where knowledge mobilisation can be drawn from real lived experiences. We can learn from communities and use these lessons in our professional circles as well as other communities or projects. We need to promote the culture of community-based knowledge mobilisation and documentation; and not only rely on desk-top based knowledge approaches.

Alois Sibaningi Baleni is a Social Scientist with experience in the development sector. The depth of poverty particularly in rural communities of Southern Africa influenced his interest into the field of Development Studies and Sociology in an attempt to generate a better understanding of the intersectionality of the social struggles of poor and marginalised rural communities. The combination of his lived experience and his ethnographic research approaches has confronted him with the realities of the narrative of the deeper perpetual existence of coloniality and the false imagination of modernity as a global solution to inequality. In his professional work he has project managed a couple of national and regional Non- Governmental funded development projects which has included projects in Zimbabwe, Mozambique, Botswana, Swaziland, Lesotho, Malawi and South Africa. He holds an impressive collection of professional and vocational diplomas, an MA in Development Studies and is currently awaiting examination for his Doctorate in Sociology looking at the intersectionality of the social cleavages of race, gender and citizenship during a violent student social movement protest of FeesMustFall during 2015/2016.

Chapter 10
Empowering Local Institutions/Communities Through Development Practices

Steve Kemp

Introduction

I arrived in Kenya, at the age of 23. That was immediately after I had got my first degree. I was certainly naïve, and I focused only on the one technical task that I was assigned. Since then, the bulk of my career has been in and around East Africa. A long learning curve, punctuated by a few 'aha moments', characterises my life.

I have enjoyed my time in the 'developing world' and learned a lot. In the process I have come across many well-intentioned people who achieve good things. In agricultural development, there is an important move to actively engage farmers as collaborators in the development of solutions. This might seem obvious. However, until recently innovations in livestock farming were purely technical in nature. We assumed that we would invent a series of silver bullets which we would hand over to farmers to transform their lives. I don't think that was wasted effort because pure science targeting African problems is a key part of the toolbox. However, it is important to remember that there are many other points of intervention before we can achieve the outcomes that we seek. One critical question relates to who should do the work that straddles science, socio economics and pure on-the-ground development. There is a real risk that large international non-governmental organisations (NGOs) will follow the money into downstream development and become a threat to the national systems that are in-turn gaining upstream capability. In the following sections, I share my personal experiential stories as I navigated my life in development practice.

S. Kemp (✉)
University of Edinburgh, Edinburgh, United Kingdom
e-mail: steve@azizi.link

T. Madzivhandila et al. (eds.), *Development Practice in Eastern and Southern Africa*,
https://doi.org/10.1007/978-3-030-91131-7_10

Mobile Technology Easing Engagement with Farmers

The emergence of mobile technology has provided an obvious mechanism to engage farmers as research partners and beneficiaries. However, the opportunity is being exploited in a rather fragmented way. Even within the same donor organisation, some see digital tools as a silly distraction from the basics. Others see it as something that should be 'left to the professionals' rather than being re-invented on a case-by-case basis to ensure alignment with local context. There are yet others who see it as a playground for new toys and magical solutions from established big companies. Clearly, all these simplistic opinions are wrong in themselves and neglect the fundamentals. But who is best equipped to identify the potential of digital tools to overcome real-world bottlenecks in a sustainable way?

One very exciting development is the emergence of smart young African entrepreneurs on the mobile data scene. For the first time we have a disruptive technology that requires only brains and connectivity to develop. This is an area where I see the large international organisations playing a facilitating role to provide opportunities and testbeds for local start-ups. It's an exciting prospect if the donors understand it and the international players can resist the urge to become implementers rather than facilitators. Once again, this will only be possible if the donors resist the siren call of the fly-in consultants with their flashy sales pitch, simplistic solution, no true presence on the ground and zero sustainability. Today, there are digital startups operating in East Africa with donor funding, who are directly competing with local entrepreneurs.

Weak Funder, Implementer and National Government Linkages

Perhaps my key learning point has been recognition of that deep disconnect between the triumvirate of funders, implementers and national governments. It is an arbitrary funding cycle, which forces everyone back to square one every few years, and drives and maintains these gaps. Even the very best international organisation which builds good, effective, long-term and genuinely collaborative relationships with its national partners, is often forced to throw all its achievements back into a funding bid. People who are usually smart and well meaning, but with a depressingly poor understanding of the fundamentals adjudicate the bids. I am still in shock at the prominent international consultant who convened a workshop in Africa to plan a very large investment and proceeded to populate it with expatriate *experts* plus a few token local farmers who were asked to participate in a humiliating mock quiz show. American and European invitees were then invited to imagine the needs and feelings of the farmers and then get on with the real business of the meeting . This unshakeable belief in the power of the Western way of doing things while the Africans are just there for entertainment, or as passive recipients of Western skills, remains a fundamental problem. This must be addressed if we are to allow genuine local capability to emerge from the trap of supply-driven, short-term, project-based investment.

Resentment of Expatriates

I lived and worked for many years in Africa. In everything I did, I believed that I was fair and respectful of my African colleagues. At the same time, I trusted that in turn I was respected for what I was bringing to the table. This was my perspective until I began to view the parallel health sector through the eyes of a smart African woman. That experience exposed how naïve I had been. Slowly, I have come to realise that there exists a deep resentment of expatriate 'consultants' (and perhaps expatriates as a whole) among young educated Africans, especially in the health sector; and usually for good reasons. Expatriate consultants flying into East Africa are often either ill-informed and unhelpful, or defensive of their own incomes. They can openly block African experts from becoming established. The phenomenon is more obvious in the health sector than in agriculture simply because there is much more money circulating and available to be syphoned-off.

To a large extent I put the responsibility here on African governments for failing to step-up and demand joined-up funding with a clear long-term goal of breaking the dependency on aid. Implementers collude in this, leaving the donors confused and powerless to do more than throw money at the latest hot topic with this week's snake oil.

As a result of government passivity, I see expatriate projects treating African governments and systems with complete contempt. Imagine a random African turning-up in an American hospital to take photographs of patients and health workers to create some kind of tableau of the health system, without any advanced consultation with the hospital or any kind of research approval. It's unthinkable, but that and similar outrages happen all the time in Africa with expatriate consultants. They seem to imagine that their skin colour will be a passport to bypass approval and regulatory mechanisms. Imagine an African flying into Washington and demanding to meet with the Secretary of Health to explain how their tiny 1–2-million-dollar project will change the system in his or her country. This might sound insane or improbable, but it happens all the time in reverse. Donor employees even try to summon African Ministers to their hotels because they are too busy to come to the Ministers.

At a smaller scale, I saw an 'expert' brought in from the private sector to show Africans how to manage a critical transition in the human health sector. The lady struggled to grasp the issues and the context. Inevitably, she blamed her shortcomings on the system that she was required to work in but did not understand. Eventually, employing a local consultant for just a few weeks rescued the situation and created something reportable for the donors. The very expensive expatriate flew home with pictures of lions and elephants including tales of her brave work in darkest Africa.

Despite saving the situation, the local consultant received little money or appreciation. Till today, the donor continues to believe that an expatriate consultant solved the problem.

Let me share yet another experience here. Recently, I encountered an organisation that a large international donor created with the explicit intention of developing African capacity in one aspect of human health. An American consultant with almost no subject-specific skills led this organisation initially, promising to step down and hand over leadership to Africans. Years later a group of white, self-declared Directors, with a white board of Governors, appointed by the same people runs the organisation. Africans within the organisation have no voice when it comes to strategic planning or engaging with national systems in ways that meet genuine need. The organisation essentially works by identifying a demand which can be met by flying-in a consultant and persuading local ministries to support their bid for resources to meet this need. Thus there is no true integrated planning or prioritisation of needs and the eye wateringly expensive consultants who fly-in prevent the emergence of the local talent which certainly exists. Thus, an organisation which is funded to develop African capacity, ends-up actively blocking that talent.

As a result of this sorts of behaviour, there is a deep resentment and mistrust of expatriates among African experts. I see this especially among the younger generation often mentored by expatriates and trained in a highly international environment. The latter, having gained seniority and experience, suddenly find themselves hitting an unbreakable expatriate ceiling. I see many of the smartest among them walking away from this fundamentally corrupt system to work as consultants. This may be the mechanism for change, growing a truly indigenous capability despite the vast sums of donor money being thrown at expatriate-driven interventions. Perhaps, slowly donors will come to see that they are blocking rather than enabling African capacity. Also, we can only hope that African governments will start to listen to their own people to come up with home-grown solutions.

Conclusion

The experiences I shared here reveal that the attitudes and behaviours of expatriates brought to Africa as technical experts are indeed retarding the pace and direction of development. Moreover, solutions to African challenges are frequently already embedded in various communities in African countries but they appear invisible to the donor community. The development world needs to understand and appropriately value the knowledge and capacity of local practitioners and understand that what they bring to the table is at least as valuable as an arbitrary and temporary technical skill from en expatriate. Local experts should be respected and granted space to lead the solution finding processes. National Governments in Africa must also take ownership of their science agenda and use that to drive their development agenda. As long as they allow their entire scientific and technical capability to be driven by short-term donor funding, then there is no chance of breaking aid dependency.

Steve Kemp is a Professor at the University of Edinburgh and has worked on aspects of livestock genetics, immunology and health in East Africa since he obtained his 1st degree in zoology from the University of Wales. Recently he has taken a lead in the use of modern informatics approaches to bring researchers and farmers closer together and to establish systems to capture data from the field and convert it into information that is of use both for long term research and short-term farmer support. Over the last 5 years, he has been exposed to a number of human health programs which provide a useful contrast to the way that agricultural development programs work.

Part III
Culture Dynamics, Skills Development and Science Communication

Chapter 11
Gender and Power Dynamics

Lin Cassidy

Introduction

As an academic researcher and development consultant, my role in rural development has been less about introducing interventions, and more about gathering information to inform interventions. This means that no matter how 'participatory' the methods I have used, my engagement has primarily been 'extractive'. My experiences have been coloured by the uncomfortable awareness of how my privileged circumstances contrast with the basic subsistence conditions of those I collect information from, resulting in the disquieting realisation that the flow of benefits is not immediately in the direction in which it is most needed.

The experiences and reflections I relate here are intended to highlight the need to be alert to the imbalances in development and the development research process. The stories are not intended to provide answers to questions but are meant to provoke some reflections on the values we bring to our rural development practices.

Being 'Other'

As if my skin colour did not already mark me as "other", I am now seated on a chair, next to the chief. All the other women are sitting on the ground, legs stretched out straight in front, while the men sit on benches and chairs facing the chief and other village leaders—and me. I pull my skirt even lower over my knees. The *kgotla*, or communal meeting, is often upheld as a place of consultation and proof of democracy. Yet women rarely speak. I wonder to myself: "Do I serve as an example of a different role for women, or am I just too different for the women sitting in this *kgotla* to relate to?" They are, after all, older women—those 'free' to attend because they do not have

L. Cassidy (✉)
Okavango Research Institute, University of Botswana, P/Bag 285, Sexaxa Road, Maun, Botswana

T. Madzivhandila et al. (eds.), *Development Practice in Eastern and Southern Africa*,
https://doi.org/10.1007/978-3-030-91131-7_11

formal jobs, those clinging, perhaps, to tradition. The chief introduces me, and in the translation, I am turned into a 'school child' because I am working in academia. Another difference highlighted, and the gulf, which I will need to cross to connect with these women later, widens.

In these moments my privileges do not help me; I clearly do not belong. How then do I leverage my 'otherness' to promote positive change without imposing it? How can I as an outside researcher and development worker forge the common bonds needed to engender trust so that I can act as a conduit for women's voices?

The truth is, women do have a voice, but so often it is drowned out by a more masculine agenda. They have informed, reasoned opinions. They have discussions among themselves and with their menfolk at home. But they do not have a way to channel their knowledge and views so that it is formally listened to and acknowledged. At the *kgotla*, we have learned that it helps to hold a separate meeting with women first, to give them time to develop a consensus position on issues, concerns and choices. In such preparatory meetings, I can join them at ground level (literally and figuratively), and dissolve into the role of collator and clarifier. In this role, it doesn't matter that I don't belong, and my otherness gives validation to the women's bold ideas. My otherness is to one side, no longer a disruption, but a tool.

Working with men and women separately is one of the simplest ways to support women from the inside of a community, without telling its members how things could or should be different. The pictures that women and men create about their community's landscape, resources, history, issues and priorities are always different. Yet there are very often points of overlap—because ultimately women and men go back to shared households. Creating gendered focus groups allows the differences and commonalities to be highlighted for all community members. It is rewarding to see how a collective position is valued by both men and women. When women and men are able to identify points of convergence in the assessments of problems and priorities, it provides validity to those aspects that are not shared. In these moments, 'The Women's View' is reified and can be respected even when 'a woman's view' would otherwise be overlooked.

The Subtleties and Scales of Gender Power Dynamics

I love statistics. There is something satisfying when patterns emerge, and abstract trends and processes are revealed. There is a wealth of information embedded in comprehensive datasets. Yet every time I do a quantitative household survey, I feel an enormous sense of loss. It starts with slow attrition during the enumeration days. In the evenings, at a camping table lit by a lamp connected to the pickup's battery, as we do our first field checks of the survey instruments completed that day, the dehumanising begins. I look at the forms, and notice how the faces, colours, textures and nuances of family life are already starting to fade. "Was this the household where the woman had just lost her husband, or the one where there was a bright red bougainvillea planted in the yard?" The individuality—whether personal fears or quirky sense of humour—doesn't have a Likert scale or box to check in our survey instruments.

By the time data entry begins, the dehumanising begins to feel like a little death. While nights in the field are sleepless because I have witnessed such poverty and been exposed to much raw suffering, nights back at home are sleepless because I can feel the characters being leached away as they are replaced by numbers during the data entry process. Where there were wrinkles and broken shoes and laughter, there are now digits—cold and devoid of personality. I feel like an executioner—killing off the true essence of what made up life for the people we interviewed, and I feel distaste that this is what I've chosen to do—turning human beings into the smallest of things—a data point. Because this is what we're after—the hard, irrefutable data is the prize. Yet scientific research can carry an emotional price.

The dataset demands so much attention—cleaning, cross-checking, calculating composite variables—that by the time it is ready for analysis, only a few of the more candid survey interactions remain etched in my memory, and there are certainly no records of them in the database. The people and the numbers are completely severed from each other. Perhaps that is how it should be—anonymity protects privacy. Datasets can feel so rich and so empty at the same time.

How then to put the data to work, to get them to reveal nuggets of information that can make up for the qualitative realities erased for the sake of empirical evidence? One of the first important understandings of gender dynamics in rural communities that I came to as a younger researcher, emerged from statistical analyses. I learned that gendered status and power dynamics are not simply a matter of whether one is male or female. This lesson came because I *didn't* sever the connection between memory and quantitative record for one household. The respondent for the household was not the head—but she was a woman in an important leadership role in her village. While in the field, I kept puzzling over what factors or attributes may have contributed to her gaining a position of influence, where other women with similar levels of education had not. Back in the office, I did not set out to mine the data for the answer. I went back to analysing households, which were the level of analysis of the study.

The analyses revealed that male-headed households were most likely to have family members in positions of leadership. What mattered was not the gender of the person in that position, but the gender of the head of the household that the person came from. Comparing my 'leadership woman' against others showed that this was the main attribute in which they differed. There were other factors, of course. Size of household, and wealth—but all of these were differences at the household level, and not the level of the individual.

Seeing such patterns in my own data, instead of reading about them, made the issue so much more real for me. I could understand first-hand just how hidden gender issues can be, and how much feedback there is in the factors influencing access to status and power. This was a story I could not see on the ground, but one which profoundly influences the recommendations for the level at which interventions should be targeted.

Discomforts and Development

Once again, I find myself pulling my skirt lower. I hate wearing skirts, but I'm not here to feel comfortable, I'm here to gather data. Once again, I'm sitting on the best chair, and the only other woman is on the ground. We're in the woman's yard. She is head of her household, and the yard is full of men—sitting on broken plastic chairs and upturned buckets. The men are drunk, and as an outsider—a white woman—I am attracting attention that makes me more uncomfortable than wearing a skirt. I am here because the random sample dictated it. This household, and not the one next door, not the tidy one which is not a shebeen, is part of the count that will give statistical validity to our study. Yet as I sit among the leers and the listlessness, the importance of statistically accurate representation has shrunk significantly. The woman is half-drunk too. I speculate that this is perhaps how she copes—with her circumstances, with her poverty, with the limited options she has available to make a livelihood, and with the kind of behaviour she must tolerate from her customers.

I wonder if I should try and find a better time to come back. But my time in this village is limited, and I have a feeling there wouldn't be a 'better time'. Not this week. A small baby, unable to walk yet, is crawling through the discarded rubbish and the sand. He is naked except for the red string around his waist offering spiritual protection, and the flies around his eyes. He crawls over to the empty food can that one of the men was drinking from but has dropped. He picks up the can, and drinks. No-one stops him.

My clipboard and extractive questions will further leach this woman's spirit. Yet at this moment, they give me refuge from the squirming discomfort of the situation. They provide an escape until I reach the point in the questionnaire with the list of assets that I can clearly see she does not have, but which—again for statistical validity—I nevertheless need to run through. I feel hot shame at how my questions highlight how little this woman has, as if I am rubbing her nose in her poverty. She laughs at some of the suggested assets—car, TV, microwave—but not because she finds it funny.

Statistics are not enough. This woman and the quiet desperation of her life will be reduced to a set of coded values assigned to a record in a database. I cannot get her out of my mind. I don't want to know that there are people as poor as this, particularly when I can give her nothing. We tell respondents, trying desperately to persuade ourselves, that while there are no direct benefits from this survey, it will come back to help them, being used to inform policy and bring better developments. These are hollow words echoing in a hollow void of need. She knows it, and so do I. This woman doesn't need more appropriate policies, she needs direct benefits, right now, today. Who am I to talk about discomfort?

"I Want to Marry You"

One of the defining characteristics of my fieldwork experiences in rural communities, is the number of times a man in the community, usually hanging out on the periphery of the action, would say he wanted to marry me. Mostly, of course, when I was younger. Almost all of these 'proposals' occurred within ten minutes of my arrival. Clearly it is not my scintillating personality or my fashion sense that was driving this attention. I used to spend a lot of time wondering what it was that elicited these offers. Was it even marriage they were talking about, or was it just sex? Was this how all women in this community were propositioned, or was it my otherness? Without fail, this interaction would place a spotlight on the gulf between the abstract, generalised knowledge of rural development I hold, and the cultural specifics of daily life in a community. If I was ignorant of 'marriage' being a euphemism for sex, this was probably for the best, because within my own culture such a proposition in a work setting would be beyond an affront. Perhaps the innuendo was intentionally offensive—a way of pushing back against someone who was outside, or a reaction to a woman who was acting in a role that threatened male dominance in the community.

While the statements were no doubt flicked out into public space without much thought, they always pulled me up short. Yet being offended wasn't going to help me accomplish my tasks in this community. Prone to over-thinking, I would wonder, if the men were really suggesting marriage, what were they seeing that made them think I was worth spending the rest of their life with? Did they see a sugar mommy, a ticket out of the limited opportunities of their rural life?

These ruminations, while on a personal and somewhat trivial aspect, have some broader value. They highlight how important it is not to assume my values are shared by the people I'm working with. By exploring how expectations within a marriage might look like to a rural man compared to how I—western-educated, feminist, opinionated—see it, I am forced to confront the fact that a proposed intervention will likely be valued differently too. I know this academically, but it is good to be reminded first-hand.

Invisibilities 1: Work and Rest

Unless one is from, or embedded in, a community, there will always be things we do not see. Socio-economic-cultural context matters when introducing development interventions. We all know of examples of technological interventions that have failed, or have worsened conditions on the ground, because they were introduced by outsiders whose personal values and cultural background blinded them to local realities. One of the most important cultural biases that workers from the Global North bring is the separation of work and rest. We assume such divisions are real in all societies. So we bring fuel-efficient stoves to save the environment and reduce the risk of women and children burning themselves on open wood fires. But we don't realise that by taking away the fire, we are also taking away the fireside, the quiet evening light and space where people unwind at the end of the day. We bring in hand-pumped wells to the middle of the village, so women don't have to walk so far to fetch water. We don't realise that by taking away the walk, we are also taking away women's opportunity for socialising, and for talking freely beyond the earshot of men.

Women's work is never done, and never more so than in patriarchal communities where women often carry both productive and reproductive roles. They must produce food, and they must maintain the home and family. Women need to undertake strategic tasks that allow them to maximise opportunities for engagement outside the household, and to pace themselves through easier tasks to avoid having additional work foisted on them. A colleague shared the following paraphrased story of her young and eager self, who went with some women to their rice fields. She looked at how time-consuming it was for the women to work their way through the field that had been planted in a very disorderly way. Excited that she had so much knowledge to bring to these women, she told them how if they planted the rice in straight rows, they could halve the time they spent in the field. And the women replied, "We know ...".

Invisibilities 2: Resource Scarcities and Commodification

Middlemen are almost always men. In interviews in a village where several women make money weaving and selling baskets, I am told how, when populations were lower, women didn't need men's help to collect plant resources. Basket-weavers would head out for day trips to cut palm leaves, grass and dye-bark. They would walk far but would usually be able to make it home in one day. Firewood could be found within an hour's walk of the village. Now, with increasing demand putting pressure on resources, collection takes more and more time. There's a point where the harvesting area is too far for women to go on foot, or too far to make it in one day. Enter the middleman. Sometimes the middleman is a household member, able to drive the donkey cart to places further away and come back with wood to last a month. Sometimes the middleman is an entrepreneur, even an outsider, who sees a need and goes out and harvests, bringing back leaves and bark to trade or sell. Both examples may appear to shift some of the burden off women, but there are other, hidden effects. Even where the man taking on collection of 'women's' resources is from the same household, there is a subtle power shift. Women become more dependent on, and even indebted to, their menfolk.

Once a resource has a cash value, women must have access to cash. With basket-making, there are a lot of ramifications. Middlemen have less interest in sustainable harvesting practices, such as ensuring the heart of the palm is undamaged for future growth, or not ring-barking an entire tree. Unsustainable harvesting can be good for the middleman, as it makes resources scarcer, and pushes their cash value higher. Basket-weavers have no control over the availability of the input resources anymore, making this livelihood activity risky. Baskets lose value as household utensils and increase value as curios. This changes traditional practices, and gives women access to cash, which can give women increased autonomy. It is easy to see resource depletion as an equal loss for a community, but the subtleties of how resource scarcity is responded to can affect community members in many different ways.

Women's Rights—Individual or Communal?

There are two of us from Botswana sitting in the annual gender workshop, held at a university based in the United States of America (USA), for graduate students preparing for research fieldwork in various countries in the Global South. There is a lot of energy and passion in the room; the attendees are all here because they want to make a difference. There is strong rhetoric, particularly among the young American women who have had little direct experience of gender discrimination in their daily lives. That women would accept a position subordinate to men is simply unfathomable and unacceptable to them. As the discussion moves to women's rights and human rights, I notice how the narratives centre on individuals. Statements such as "*I* would never tolerate that" and "*I* would change this" lead to discussions of how a woman should respond to inequalities in power and decision-making. What is striking is that none of the students in the room conceived of human rights as other than individual rights. Even I, with Northern ancestry and cultural roots, am taken aback by what feels like an egoistic selfishness. I turn to my Botswana colleague, and ask her how she views human rights. She tells me that growing up, her grandmother used to talk about what was good for the community, not what was good for a person. That when people looked at costs and benefits, they looked at what this meant for the whole family, and in some contexts, for the village. They did not spend a lot of time thinking about whether they had 'human' or 'women's' rights. There were the rights and responsibilities that everyone had as part of the community, but these were rarely reified or concretised. We chat for a while about the pros and cons of each of these different positions—the highly self-oriented focus, or the self-effacing one. Do women in Africa struggle for equality precisely because they are more likely to focus on collective needs—what was good for the household—than on their individual needs? Is it more 'moral' to think of the group before one thinks of oneself? Ethics aside, it is also possible to explore reasons why different cultures focus on human rights at different levels. There are stories to be told about traditional rights, and the rights of traditions. Perhaps it is only when a group's identity or existence is no longer under threat that the focus can shift to a more individual level.

In rural development, individual rights are often promoted due to the value systems in the (Global North) host countries of most donor agencies, and without regard to the perspective of individual versus collective rights held in any given community. How often do you see an African anthropologist studying European societies? I wonder how the switched contexts would influence interpretation of values on the ground.

Conclusion

It is hard for our lessons to stay current. In an increasingly connected world, social and economic values are changing faster than ever. At the same time, globalisation reaches the urban cores of developing nations far faster than it reaches the peripheries, increasing the economic and cultural gap between urban and rural areas. It is hard for outsiders to know what 'traditional practices' look like anymore, and hard to understand at what point of a development trajectory any given community may be. What is important is for development practitioners to be ready to learn and adapt, and to be open to surprises. An important tendency to fight against is confirmation bias—we know what we expect to see, so that's what we look for, and that's the trend our eyes will pick out of the data.

Women's roles are shifting too. While female-headed households are increasing, this does not automatically mean that women are gaining equality in terms of political or economic status. What is certain is that women in rural communities have a lot to teach us, if only we take the time to listen more closely. Our need to listen is not solely to ensure that the interventions we propose are more appropriate and cognisant of gendered differences, but also to learn for the sake of our own personal wisdom and understanding of the world.

Lin Cassidy started her career working with rural development and has lived and worked in Botswana for over 30 years. At present, she works as a land change scientist focusing on landscape processes, social-ecological systems, natural resources management and human-environment interactions in savannas and wetlands—still focusing on rural areas. She is an independent researcher and freelance consultant, having previously worked at University of Botswana's Okavango Research Institute, where she currently holds adjunct faculty status. Her career expertise includes land use planning, socio-economic and ecological surveys and assessments relating to conservation and development. She has a Ph.D. in Geography with Interdisciplinary Certificate in GIS and minor in Political Ecology, and a M.Sc. is in Interdisciplinary Ecology.

Chapter 12
From Calling to Collaboration: God Calls, God Empowers and God Teaches

Sue Walker

Learning to Respect Local Customs in Mpumalanga Province, South Africa

On an expedition to Mpumalanga to explore the value and range of types of kraal manure stored on farms across a farming area, a younger researcher, an extension officer and I were visiting remote areas near to the edge of Kruger Park, northeast of Bushbuckridge. We drove on many poor gravel roads to visit the villages. After greeting the head of the household and explaining our purpose via a translator, we proceeded to the kraal where the cattle were kept overnight. As I was about to enter by climbing through the poled enclosure, the shouts of bystanders made me stop in my tracks. I could not understand this, as we had requested and obtained permission to collect the dung samples. Why suddenly was I stopped in mid-stride before my feet touched the ground inside the kraal? I tried to think quickly—was there some danger? There were no cattle in the kraal during the day so no harm could come to me, so what was the problem? After a heated discussion between the farmer and the extension officer, that I could not understand, I waited patiently for them to decide. It appeared that the head of the household said they could never imagine that this 'umlungu', white, educated 'udokotela' woman, was actually going to dig in the kraal herself! This is not allowed. It would be a bad omen on the fertility of their cows. No women of child-bearing age are allowed in the kraal even if the cattle are not present. Oh well, so I could not be hands-on with the collection of these samples, and my colleagues had to do the digging while I stood and watched!

S. Walker (✉)
Agrometeorology, Agricultural Research Council—Soil, Climate and Water, Pretoria, South Africa
e-mail: walkers@arc.agric.za

Department of Soil, Crop and Climate Sciences, University of the Free State, Bloemfontein, South Africa

T. Madzivhandila et al. (eds.), *Development Practice in Eastern and Southern Africa*,
https://doi.org/10.1007/978-3-030-91131-7_12

That day I learnt that even where we may have explained our purpose and procedure from our perspective, perhaps from another viewpoint and culture, we may be violating local customs, values and beliefs. We need to approach things more slowly, explain each step and exactly what each person will do before beginning and unintentionally violating customs and traditions. A lesson well learnt early in my career. This example emphasizes the need for researchers to be always unassuming and ready to align quickly with local values, customs and traditions. One should do background preparation on the local customs and behaviors before visiting a new area. Although this story may sound discriminatory against women of certain childbearing age, it also shows that some cultural norms and values could protect women from engaging in some seemingly strenuous endeavor, or dangerous work with livestock. It shows that in development work, one needs to be ready to respect the local values and be willing to change ones' own attitudes.

Giving Gifts in Ethiopia

I remember in the early days of my rural village visits, I would always try not to go empty handed as I love giving gifts and I therefore carried a bag containing ballpoint pens to hand out to the school children and others. When engaging with the Ethiopian children along the road they would call out "Bafana Bafana" (the name of RSA national soccer team) when they heard I was from South Africa. Or they would want to talk about "Madiba" (Nelson Mandela) soon after his release from prison to become South Africa's first democratically elected president. He really was a hero throughout Africa and not only for South Africans. Through this I learnt that people were interested in my country and I needed to learn some background about theirs prior to my visiting them. This shows that one respects their country when one is doing development work in a foreign country.

On other occasions I would take University of the Free State (UFS) caps with me as gifts for my counterparts. This practice began at the end of one trip when I gave the driver, who had taken me around all week, the UFS cap that I had been wearing. He was so thankful. So on subsequent trips in Africa I always tried to take a few extra UFS caps with me. When I was leaving I would give them to the students and their colleagues, telling them 'This cap needs to be returned to Bloemfontein, so I hope you will visit us!'. Many of these students did in fact come to Bloemfontein to further their studies (Fantaye, 2004). Sometimes the problem was that the caps could not expand to be large enough to fit on their heads. For example, when the ladies had a large curly wig or hijab or head scarf, it was difficult to balance the cap on top! Once a visiting professor from The Netherlands struggled to keep the cap on his head so he passed the gift on to one of the extension officers.

It is well-know that traditionally African mothers teach their children to 'ukuphaphatha' give a few gentle claps of the hands before receiving a gift. I often experienced this tradition when handing out gifts. So it is not only the Japanese that have a tradition of gift giving, it is also a common practice in Africa. My joy of giving gifts was reinforced as I saw the effect is had on people and especially when I realized that the gift of a pen to a child often enabled them to learn to write. Many times in development projects, it would seem that the researchers only come to collect information and take something of value for themselves in a selfish manner, when the purpose of many of the development project is to uplift and capacitate the local community members. If one can at least bring something that is a practical gift or need of the community when arriving and being introduced, then it shows a different attitude in one's heart. So the gesture of bring small gifts of pens for the children helps to break the ice and show one's good intentions. If one understands some of the perspective of local traditions or value system and can show goodwill and acknowledgment of the hierarchies and etiquette, then the introductions can go more smoothly.

Learning Humility and Gratitude in Zimbabwe

I was often taught humility when visiting villagers. On one visit, in the middle of a drought in Gwanda, southern Zimbabwe, I remember our hosts calling my colleague and asking us to bring salt along (Mupangwa, 2008). From my previous experiences in South Africa, I knew that the farmers called fertiliser 'salt' or 'sugar' (ushukela) which had me confused—surely they would not sprinkle salt or sugar next to their tomato plants? Did the Zimbabwe farmers also wanted us to bring fertiliser? However, when we visited the farmers, the whole story was explained: By the end of the week, the soup is so watery and tasteless, as they prepare a big pot of soup over the weekend, and keep adding water every day. So if they could add some salt to the pot, it would at least give the watery soup some taste. I learnt that some ladies have a solution to really stretch the vegetables but I also learnt how hard life is during drought periods.

During that same trip we found the farmers so generous with the little they had. One lady ran to the tree and brought back the small muslin cloth containing sour milk to share with us. Others invited us to share their meal with them. Later, we always stopped at the shop in town to buy as many as 20 loaves of bread to take with us each day. At least we were offering something more than our agricultural expertise when we met the farmers, as the scripture says 'man does not live by bread alone' (Matthew 4:4).

Another time I walked to the long-drop toilet at the edge of the compound and was surprised to find a beautiful, shiny floor, obviously polished with cattle dung every day. This reminded me of a high school headmaster who said that if you want to know what type of people they are, go their bathroom and see how clean it is or how they keep it. So the farmer even took great pride in keeping her outhouse pit latrine in a sparkling top condition for her own family. This shows character! That day I learnt from this farmer that it shows when we take pride in even the smallest things. This shows how when working in development project, we need to respect those we are working with and engage about simple everyday issues as we go about our project related work. Realizing that we can learn from each other as we conduct the farm visits and focus group research (Mupangwa, 2008).

Learning to Appreciate Other Cultures in the Butana Region of Sudan

During my trips to the Butana region of Sudan, between the two Nile rivers (i.e. south of their junction in Khartoum), we had a number of unique experiences during which I was learning in leaps and bounds—new things and ways of relating to rural people every day. The Sundanese student was doing a project on using remote sensing to determine the impact of desertification (Elhag, 2006). During the ground-truthing exercises we travelled through the study area to visit several farmers across the area and assess the state of the land to compare with the satellite images. Some of lessons learnt were about respecting the cultural values held by the local farmers—for example, do not ask how large the farmer's flock of sheep or goats is as there are some superstitions about never counting the sheep as it brings bad luck or a curse to the farmer. Or, perhaps they are afraid that someone will report them to the taxman and they will have to pay taxes per head of sheep to the government.

One day my student slipped away to pray while we were talking to the farmer. She was so surprised on her return that I had been able to hold a conversation with the farmer all by myself as we did not speak each other's languages. We discussed the start of the rainy season, the growth of the wheat crop and collection of water from the overland flow into a man-made structure (called spate irrigation). We communicated by pointing and with gestures so we could connect in spite of the language barrier. Sometimes the lack of language can be a major barrier in development projects. However, non-verbal communication is a special skill that comes with experience of being in widely varying countries and cultures. Such skills include transmission of messages or signals through a nonverbal platform such as eye contact, facial expressions, gestures, posture, and the distance between two individuals. In this case without language skills, it included facial expressions, body movements and gestures to explain some aspects of weather phenomenon and farming activities. These could be considered as examples of 'snowball sampling' as the full extent of the population is unknown and widely distributed, so it is difficult to choose the subjects or to assemble them. As one moves through the farmlands, one will encounter the farmers in the fields and can then engage them. This farmer will also then refer the researcher to the neighbor and so the sampling can snowball and the sample size is increased through the qualitative research methods interventions.

On another trip, I had the amazing experience of travelling across the Nile River into the deep rural desert area with one of the leaders of the community. In addition to that we met the man's three wives at different homes all on one day, starting in town at the compound where he ran his transport business. We met the young wife, number three, who entertained us with a cup of coffee as we sat on a luxurious and plush modern couch in the front room of her house where we played with their 6-month old baby. Leaving town, we drove towards the Nile River and waited our turn to cross on the punt or motorised floating raft. I enjoyed watching the herons and other water birds as we waited. As we drove along, he explained that we would stop at a school to greet his second wife who was a teacher. I hope I did not sound too surprised. She was smartly dressed, sophisticated and could speak good English, so we could easily connect with her. We also stopped at the little shop and met some of the other locals. We then proceeded on our way further into the Butana desert. Slowly we saw fewer and fewer trees and then hardly even a bush. Eventually, we arrived at a small windswept village with sand everywhere, half broken mud walls around the compounds and houses constructed with mud bricks. The entrance to the compound had walls on both sides, piled high with wind-blown sand. There were two or three houses inside the compound and many young children ran to meet us. Then we met his first wife, a thin, bent, elderly looking lady with a big smile, but with sun-dried skin and hardworking hands. Lunch was served, and we were allowed to eat in the 'men's house' despite being two women, as due to our education status we were considered 'honorary males'. Following a delicious lamb stew, we had a siesta, again on some day beds in the 'men's house'. As it cooled down later in the afternoon, we began our tour of the farming enterprise, walking through the dry stony lands, which looked almost like a gravel road! The farmer stopped to show us the tiny new grass sprouts—each only 1 to 2 inches long and spaced far apart, struggling through the hard stony ground. We arrived at a well where sheep and goats were drinking water from a trough. (I did not ask how many there were as I had learnt my lesson already). Contrary to the way we normally build a dam in a valley, this dam was built on level ground. It consisted of three high mud banks with an opening on one side to allow water from overland flow of the last tropical thunderstorm to accumulate. So I could see that one really needed to do good environmental analysis before recommending infrastructure changes under this harsh desert area (Elhag, 2006).

What a day of new experiences and lessons! It was the first time I had met in one day three wives of one farmer who lived completely different lives. Such cheerfulness and humbleness of wife number one, living in the deep rural area far from a town, yet a stable historic homestead like an oasis in the desert. A hard worker supporting her extended family of several generations—in contrast with wife number three, living in luxury in town with a young baby. As an outsider, I do not judge them but accept their way of life and beliefs as they live their traditional lifestyle with their religious values. One has to mature in ones' outlook of acceptance of others cultures, religions and languages to be able to work in cross-cultural situation.

Realising Ones Contribution Impacts Many Lives Near Monze, Southern Province, Zambia

One of my students worked in the Mujika villages near Monze in the Southern Province of Zambia, and I visited a number of times over a period of four to five years, so I was privileged to revisit several villages and gather fond memories while learning my development lessons (Nanja, 2010).

Each year we shared the seasonal outlook for rainfall with them, as was our routine practice. As we predicted good rainfall we advised them to plant a long-season variety. We then started to brainstorm what opportunities they have to make some money in order to buy the long-season maize seed. One of the actions was to collect reeds from near the river for crafting mats or baskets. These products could be sold at the nearby markets or in town. The next day, as we drove through the farmers' fields, a young child came running towards our vehicle shouting—"stop – wait"! As we stopped and got out to greet him, a lady came running across the field with some parcels in her hands. She said:"Remember yesterday, you said we could sell mats to get money for maize seed? So here are my mats. Please buy them now". Well, what choice did I have? I got out my purse and negotiated a price to purchase the mats. A good lesson in 'practice what you preach' or 'be willing to be part of the action and solution for the farmers'. As it is explained in the scripture "Therefore, you should treat people in the same way that you want people to treat you" (Matthew 7:12). When you have a person-centered culture, then it promotes development through participation, empowerment and learning with the leadership reflecting and being an example of these principles. As leaders, we also have a responsibility to be role models of what we are teaching and build learning cultures around us. Being a role model shows our integrity and that we live by the principles that we promote and advocate no matter where we are.

On another occasion, we were diligently trying to assess the needs of each village in a participatory fashion together with the community, working out some ways for the headman to address them. The community members acknowledged certain barriers to understanding the seasonal rainfall forecasts and other advisories about changes in agronomic practices. Even if we gave them posters or flyers with climate information, they could not read it, revealing that most of the female farmers older than 30 or 40 years could not read and write. We discussed back and forth how this could be addressed by some remedial action. Suddenly, one of the braver elderly farmers said that her granddaughter can read and write! So, we led them on to ask how this could be applied to solve the problem. To cut a long story short, two young ladies, one from each village, each with a baby on their back, and who had about 8–10 years of formal schooling, agreed to be 'teachers' for the elderly ladies. We established two small reading classes or a school that same day! I work in agriculture, but when the farmers needed to learn to read and write, we had to help them with this transferable skill. So the next day we brought some exercise books and pens from town for them to use, and a local colleague visited and encouraged them through the next few months. In actual fact these villagers came up with the solution from within their own community, but needed our support and encouragement to implement it. A few months later, when I visited the village again, as was the habit, we were greeted at the edge of the houses and were escorted to the meeting place under the big tree. As we danced and sang together walking to the meeting place, one of the ladies came up close to me with big eyes and a huge smile, and said in perfect English "Good morning, how are you today?" Wow! We were able to communicate for the first time without the use of a translator! They had not only learnt to read and write their mother tongue, but also to speak English. I learnt a lesson that day, and was a beneficiary myself in an unexpected way how a little help and enthusiasm as a catalyst can stimulate and support a community. Two classes had been formed and several ladies became literate during that year. This changed their lives and gave them self-confidence and opportunities for trade and communication. But there was an unexpected kick-back for us as researchers as the window opened to communicate directly with the farmers. Even though my role was to develop and encourage more sustainable agricultural practices, in that development situation, I needed to also assist with the development of basic transferable skills.

We learnt more life lessons in the meetings held to spread the seasonal forecasts. At another village meeting, attended mostly by women and very few men, the group decision was to not plant on the lowlands (being mostly heavy clay soils), but rather to plant on the uplands with more coarse gravelly soils, due to the above-normal rainfall that was expected that season (Nanja, 2010). Upon our return near the end of the maize growing season to inspect the on-farm trials, we held a farmers' day to walk through the fields and discuss the crop results. During one meeting, I saw a man standing far off and not joining in the group discussions with the rest. When I enquired about our omission to invite him to join us, the villagers explained that his wife was at the pre-season meeting and she wanted to plant on the uplands, but he refused to listen to the advice from the 'witch from outside'. So his field was standing in pools of water for the last month and the maize had turned yellow, this despite the good rains or actually because of the good rains. He had a harvest failure and was too embarrassed to attend the meeting, as it felt like salt being rubbed into his wounds. This goes to show that we need all family members, and especially the head of the household, to attend the meetings or else the local traditional hierarchy and decision makers may not accept the scientific advice for a change in the best practices. Again there was a reminder that when working on development projects, one needs to involve the decision makers as well as the workers in the transfer and communication actions.

Conclusion

These experiences point us as researchers to an enriched understanding when we venture out to visit and engage the farmers by walking in their fields. Researchers should acknowledge and appreciate the farmers' culture, current activities and desires or dreams for their future sustainable livelihoods and continued production of food.

We need to give them freedom to use their own initiative in making changes or adapting their farming practices. We learn to appreciate the value of Ubuntu in action and as role models by our actions, resulting in improved scientific research outputs. I have grown and expanded my outlook and know that God calls, God empowers and God teaches, so that we can develop a future together.

Acknowledgements I could not have visited all these farmers without the support of the PhD students by many different funding streams. I thank the PhD graduates—Dr Kindie Tesfaye from Ethiopia, Dr Walter Mupangwa from Zimbabwe, Dr Muna Elhag from Sudan and Dr Durton Nanja from Zambia.

References

Elhag, M. M. (2006). *Causes and impact of desertification in Butana area of Sudan.* Ph.D. in Agrometeorology, University of the Free State.
Fantaye, K. T. (2004). *Field comparisons of resource utilization and productivity of three grain legume species under water stress.* Ph.D. in Agrometeorology / Agronomy Interdisciplinary, University of the Free State.
Mupangwa, W. T. (2008). *Water and Nitrogen management for risk mitigation in semi-arid cropping systems.* Ph.D. in Agrometeorology/Agronomy Interdisciplinary, University of the Free State.
Nanja, D. H. (2010). *Dissemination of climate information to small-holder farmers: A case study for Mujika Area, Zambia.* Ph.D. in Agrometeorology, University of the Free State.
Walker, S. (2020). Value-added weather advisories for small-scale farmers in South Africa delivered via mobile apps. *Irrigation and Drainage, 2020*(Special Issue), 1–7. https://doi.org/10.1002/ird.2506

Sue Walker From a young age I believed that God called me to 'feed the people'. He opened many doors enabling me to hone my skills in order to fulfil that calling. Because of this I call myself an agrometeorologist with a heart to serve the people. I always enjoy sitting in the dirt with the farmers discussing their land, their crops, working methods, how to improve production and ultimately their lifestyle. The way I see agrometeorology is that it includes the people dimension as well as the interaction between farming and the atmosphere from the leaf level through the field and farm level all the way to the regional and global scale. I have always enjoyed projects where I could interact with the farmers and I have worked hard to make agromet research applicable to on-farm decisions. Soon after attaining my bachelors degree I worked on a project to help farmers improve their productivity by optimising their irrigation, where we would visit the farms and drink coffee before walking through the field. These experiences really helped me to keep my feet on the ground (literally!) and in touch with real day-to-day farm decision making. So I began exploring participatory methods and 'farming systems research extension approaches' to develop research projects. Later, I tried to make sure that each of my PhD students devoted at least one chapter of their thesis to consider the effect of their research results on the farming community. More recently, I have been involved in co-developing climate services for the agricultural sector such as CAPES (Community Agromet Participatory Extension Service) in the Southern Province of Zambia (Nanja, 2010), and the AgriCloud mobile phone app for rainfed farmers in South Africa (Walker, 2020). I gladly share some of my experiences from my interactions with the various communities and the lessons I learnt along the way.

Chapter 13
Journalism and Communication in Development Practice

Mantoe Phakathi

Introduction

Stories always fascinate me. Thus, it does not surprise me that I ended up in journalism. Over the years, journalism has exposed me to people of different backgrounds, from the most influential politician to the farmer in the most remote rural area. I got access to many stories which in turn I shared with wider audiences. I spent most of my journalism career writing for a political magazine in the Kingdom of Eswatini, formerly Swaziland, and online news agencies. I covered all issues, but my focus was politics. Politics has an impact on almost everything under the sun. How a farmer, in any country, will perform is partly impacted by the decisions of policymakers.

Although I covered a wide array of issues through a political lens, my passion remains biased towards agriculture, the environment and climate change. With an increasing global population, food security has become the main concern of many governments across the world. Climate change impacts such as droughts and floods are reversing most gains that have been earned over the years. The situation is not different in my country, Eswatini, where droughts are destroying crops and leaving many families with no food. Despite having reported on climate change, I still find the subject very complex and technical, and many times requiring me to read and consult in order to unpack it for wider audiences. This can often be tricky, especially when one lacks the technical depth.

While I have enjoyed journalism, recently I have steered more into communication. I help development organizations to reach out to diverse audiences. Many organisations are carrying out wonderful work in development but do not tell their stories fully. My work provides a platform to expose both the impact that the institutions have on the society as well as training staff working in this space.

M. Phakathi (✉)
University of South Africa, Pretoria, South Africa

T. Madzivhandila et al. (eds.), *Development Practice in Eastern and Southern Africa*,
https://doi.org/10.1007/978-3-030-91131-7_13

Effective Communication Could Have Saved Us from Conflict

I had never felt any sense of danger while doing my work in development communication. That was until one day when I was part of a climate change project that aimed to support farmers to adopt climate-smart agriculture (CSA) techniques, linking them to markets and equipping them with business skills. The project was also remunerating workers aligned with the association of farmers to improve production because most of its members were elderly.

On this particular day, the project manager asked me to be part of the team on a project to capture the experiences of workers who had not been paid because of incomplete paperwork that farmers submitted. Being a consultant who was working remotely on this project, I was not privy to the final details of this administrative issue. My first impression was that this was a minor challenge that needed the attention of project staff.

We got to the meeting and met with the farmers. Only then did I realise that the workers had not been paid their wages because the farmers association's designated committee members had not properly completed claim forms that captured the work and time of each worker. The project office withheld payment, calling on the designated association's committee to capture information properly. The finance officer justified this with the argument that the project had to account to auditors with the right and fully completed paperwork. The meeting with the farmers was characterised by a blame game among the designated members of the association.

The workers operated on a 23 ha field. They were weeding, cutting trees and irrigating their crops. Later, they started gathering next to the shelter where we met with the farmers. They were carrying all their tools, namely hoes and bush knives amongst others. Such could easily be turned into weapons. Later, they were later called into the meeting where the Project Manager and Chairperson of the association explained why payment had been delayed. Also, they were informed that the project team would engage the farmers and facilitate processing of paperwork before release of payment. The workers rejected the explanation and proceeded to lock the gate and demanded that we would not leave before paying them. The fact that they had their tools close by made the situation even scarier. Later we learnt that the farmers had told the field workers, who were mainly young people, that the project team would bring their money. This did not happen.

It took more than two hours of negotiation with the field workers, who did not want to hear anything else but getting their money. Even though the situation was tense, the Project Manager and Chairperson managed to restore calm. They achieved this by working with the aggrieved workers and farmers to understand how to complete the forms correctly in compliance with requirements for payment. The farmers were retrained on how to complete the claim forms. After this, calm was restored. They allowed us to leave on good terms.

Through the experience articulated here we learnt that effective communication could have cleared all the challenges we faced earlier. The project was paying the field workers. Our team relied on the farmers to give them the right information, which did not happen. We learnt the need for immediate retraining of the farmers as soon we discovered that the claim forms were not completed correctly. Another lesson was that transparency is key. The workers seemed to be unclear about the project's expectations and that came with a cost. Eventually, trust was restored, and implementation of the project continued. This experience made me realise that development work can be dangerous.

Sometimes You Get Caught in the Crossfire

In 2015 the government instructed security forces to evict some residents from a piece of land that it owned. The government wanted to clear that land so that an information, communication and technology project could be introduced. The farm dwellers, as they are referred to in terms of the Eswatini Farm Dwellers Control Act of 1982, being evicted believed they were living on traditional land. They had received it from chiefs and headmen.

I was a reporter for a magazine at that time and obviously the evictions captured my interest. One day, in the company of a photographer, we searched for the evicted families after I had learnt that they were living in tents somewhere on the outskirts of Manzini, the commercial capital of Eswatini. When we reached that place, we were told that they had moved to a place 15 km away because the NGO that had lent them the tents took them back on grounds that they had to help other people whose houses had collapsed. A good Samaritan had given them space with incomplete houses to stay in.

When we arrived at their new place of residence, it seemed they were already expecting us. Never in my life as a journalist had I experienced so much hostility from people when developing a story. They were itching to physically assault us. Later, we learnt that the hostility emanated from the fact that one of the newspapers had made false reports about them. The latter newspaper had claimed that they were fighting the police and "illegally" occupied the land they were being evicted from. Having lost their homes and personal belongings and spending cold nights in tents with children without enough food, it was easy to understand the origin of their hostile reaction. After protracted negotiation and persuasion the crowd calmed down. That made us to learn that sometimes one has to call first and gauge the mood of sources rather than just showing up.

Sometimes You Get Confronted by Realities You Never Assumed When Doing Field Work

I have never considered language to be a barrier to the work I do in Eswatini. Most citizens speak siSwati while there are a few isiZulu-speakers. The latter understand siSwati. The few that do not speak siSwati are mainly expatriates who speak English. So, when working with communities in rural areas, siSwati is the dominant medium of communication. Obviously, with expatriate workers such as development workers or doctors English is the preferred language.

My perception on this matter changed one day when I was working on a food and nutrition insecurity story at a remote place called kaShewula, which is found along the border with Mozambique. KaShewula is one of the poverty-stricken areas in the country. Inadequate access to food is a major issue there. As I moved around the area on my journalism expedition, I stumbled upon two old ladies at a homestead. They spoke Shangana only but could not siSwati. There I was. How were we to communicate given that we did not have a common language that we spoke? I had to find a way of dealing with this matter.

My discussions with the old ladies' neighbours revealed that they had a granddaughter who attended the nearby school. She could be the translator because she understood siSwati. Off I went to the school in search of her. Luckily, I found her. We went back together to her home and indeed she competently helped with the translation. I learnt a simple lesson on that day. As development workers, we often make assumptions about the people we work with, especially grassroots communities. As I retreated, I wondered how many people had been disadvantaged because of assumptions we often make. How then should we reconfigure our mindsets to eliminate such thinking?

Conclusion

Clearly, the few stories I have shared reveal that even though I have spent almost 20 years in the communication and journalism space, I do not know it all. Each day is rich with new lessons ready for me, provided I am properly attuned to navigate the learning curve. Working with farmers who are working so hard to produce food for us against all odds remains one of my most enjoyable experiences. As a communication specialist, my fieldwork is always a fulfilling experience. I get to listen to their stories and lived experiences, take pictures and videos to create content which I share with broader audiences. The breath-taking scenery of my country and other countries make me look forward to my next assignments.

Mantoe Phakathi is a Swati communication specialist and a freelance journalist with experience spanning close to twenty years. She is passionate about human rights, environment, climate change, agriculture, health and other social issues. Her passion for environmental work led her to pursue an MSc in Climate Change Development and Policy at the University of Sussex in the UK in 2016/17. She is committed to supporting the climate change agenda with an interest in using communication products to make information on environment, climate change, low-carbon development and related issues to be easily accessible to all types of audiences to help influence policy making. She has a BA Degree in Communication Science from the University of South Africa and Post-graduate Diploma in Media Management from Rhodes University.

Chapter 14
Situating Self in Socially and Politically Contested Spaces

Emaculate Ingwani

Introduction

My experiences in conducting, supporting and coordinating research, and the Integrated Development Projects has brought me close to developmental studies. My research focuses mainly on land; ecosystems services and livelihoods; socio-spatial justice; sustainable human settlements; and small towns. My interrogation of 'new' urbanisms and socio-cultural imperatives in rural, urban and peri-urban zones is enabled through blending skills in social research with community engaged teaching and learning practices. My first two experiential cases show the profound effect of culture on research undertakings in rural and peri-urban communities. Whereas, the third case presents an opportunity for teaching and learning encounters through community engaged practical experiences. The fourth case demonstrates ways of dealing with unexpected practical challenges embedded in bureaucratic local government systems.

Situating Self in Socially and Politically Contested Spaces

This narrative is based on "lived" experience(s) derived from an interpretation of social and political contexts that characterize land transactions in peri-urban communal areas of Zimbabwe. By nature, peri-urban communal areas of Zimbabwe are spaces in transition from rural to urban largely characterized by a coalescence of activities (including political), structures, and people. In my effort to unravel what would become of peasant farming and household survival strategies amid the prevalence of land transactions and confusing rural development policy in these

E. Ingwani (✉)
University of Venda, Thohoyandou, South Africa
e-mail: Emaculate.Ingwani@univen.ac.za

T. Madzivhandila et al. (eds.), *Development Practice in Eastern and Southern Africa*,
https://doi.org/10.1007/978-3-030-91131-7_14

zones, I sought qualitative descriptions of phenomena through field research—to avoid 'academic tourism'. In doing so, I realized that seeking permission to undertake research in socially and politically contested zones such as peri-urban areas is bureaucratic and cumbersome; and therefore requires patience, tenacity, and humility. I adopted the top-down approach to get permission to carry out research in one of the peri-urban areas of Zimbabwe's cities. Application of the bottom-up approach through the local structures constitutive of traditional leadership could not work because the traditional leaders were not willing to allow me to carryout research for fear of reprisal by the government. By interacting with strangers, these traditional leaders could be labelled as sell-outs. As such, permission to carry out research was sought from and granted by the responsible ministry. This approval was cascaded to gain permission at provincial, district, and village levels. I thought I was sorted, but soon realized that there existed a parallel political structure of land administration. The district authorities accredited my research and gave me a letter to take to the Chief who welcomed my research idea, but referred me to his secretary, one of the Headmen in the district where the peri-urban area is located. I thought I was done with protocol and prepared to get started with fieldwork. Yet, I was fooling myself!

I needed to observe a traditional process called 'kuombera' in order to obtain audience from the Chief before commencing fieldwork—something I never planned or dreamt of. This process was critical for me to get data from the residents, as well as to walk-about in villages during fieldwork. 'Kuombera' is a Shona word, which means 'to pass greetings'. Fulfilment of the practice was through offering a valuable token to the Chief such as a live goat and 'shamhu', that is, a stick for driving the goat, or a monetary token equivalent to US$40.00 for a goat and US$10.00 for a stick. Culturally, the process of 'kuombera' and bringing gifts to the Chief signifies respect, loyalty, obedience, and submission to traditional authority. This process is not foreign, and is a prerequisite in the culture of the Shona. Contemporary 'kuombera' is thus monetized, and is very important to negotiate entry into peri-urban spaces for research purposes. I did not have a live goat. I therefore opted to pay cash—(US$50). Ideally, 'kuombera' takes place during an open ceremony at the Chief's council. During this ceremony, I crawled on my knees to the podium in order to demonstrate my submission to local tradition and authority, as well as 'going native' in research. In this regard, 'kuombera' figuratively and literary demonstrated begging not only for permission to carry out research, but the so much valuable data in a contested space. Clearly, understanding the relational sequence of tradition and custom vis-a-vis conventional and political land administration structures was critical to be vetted and to enter spaces where positionality of powerholders determine the next move. I was offered accommodation at a local school in one of the villages during fieldwork, which was designed to enable me to engage in the observation of the cultural processes and parallel systems of administration in a peri-urban space. I realized that gaining access into socially and politically contested spaces requires contextual application of social skills. As such, I assumed the status of a 'village woman'. In line with the Shona tradition, I always wore a wrapper and a head scuff throughout the field research. When approaching the yard, (usually not fenced nor gated), I asked if there were any dogs at the homestead—I fear dogs!

During each household interview, I greeted the household members in vernacular and addressed household heads, Village Heads and Headmen as 'samusha'. Alternatively, I greeted people using their totems. People really felt fulfilled, and appreciated this gesture because people belonging to the Shona tribes are generally proud of and value their totems and household headship. During interviews, I always opted to sit on the floor or ground rather than a chair or stool in order to uphold power dynamics. Shona women generally are submissive to authority. Sitting positions signify authority and status. As such, men use conventional seats or even stones to elevate themselves through seating. By opting for a lower seat (on the ground), I submitted to the household and local authority.

During interviews, respondents rambled off into political issues that were not directly relevant to my questions maybe to tap into my allegiance to the ruling or opposition parties. In some cases, the interviews offered the best platform for some interviewees to air their views or to demonstrate loyalty to the government in power or vice versa. In such instances, I redirected conversations to focus on my research questions without upsetting or taking sides with the interviewees. Some interviewees were insecure to tell their stories, and therefore did not disclose much detail during interviews. They thought I was a disguised agent from the district local administrative authority and the government. This is because land transactions in peri-urban spaces are a sensitive topic because the land under customary land tenure system is untradeable. Yet, Village Heads, Headmen, and the Chief commodified such land outside the structure of the customary land tenure system through renting and selling.

On the other hand, some interviewees did not view me as a researcher per se, but as a medium through which they could present their problems to the district local administrative authority and the government with regards to prevalent land transactions in the peri-urban communal area. In this regard, the community residents viewed me as a potential mediator who I was not. At one point, I was offered a piece of land to buy in one of the villages I was studying. When the dates for the national elections were announced, representatives of the ruling party mistook me for an agent of the main opposition party. It became difficult for some community residents to trust a stranger like me under such circumstances. This way, the community residents and the district local administrative authority officials became part of the complexities that I was studying. Although it was scary, I continued with the household interviews until I reached saturation.

A day before my departure from the fieldwork, I met one of the old women I had earlier interviewed. She was so excited to see me around, and therefore called other village women to 'come and listen to this lady… she is asking whether we want to remain rural or change to urban. I told her we do not want to change'—something I never said. I realized that some of my interview questions triggered expectations that I never thought of or able to handle as a mere student even after pretesting the research instruments. That same night, I just packed my stuff and left the village without saying goodbye. My mission had been accomplished. Clearly, situating self in socially and politically contested spaces could be dangerous and fun at the same time, but requires calculated composure.

Protocol First or Else don't Talk to Me!

This narrative captures my experience with a research team in northern Zimbabwe sometime in 2012 when we were conducting a baseline socio-economic assessment in one of the urban areas' neighbourhoods on behalf of a consulting firm. We were a team of two female and two male researchers. We were given a list of stakeholders to consult, and a well-crafted stakeholder map to reflect upon. One of our targets was the Chief of an adjacent peri-urban area. When the project team embarked on fieldwork, the assumption was that all necessary consultations and preparatory work on interviews with the stakeholders were already done, and that our role was simply to collect data. Data collection from households and political representatives such as the Ward Councillor and the Member of Parliament was rather easy because these categories of people really looked up to the benefits of the project to the local community. When it was time to meet the Chief and his tribunal, we realized that there was no prior request for permission to interview the Chief. After realizing this error in the public consultation process, we were calculative and decided to collect household data ahead of the interview with the Chief because we were likely not to do the baseline survey if the Chief did not grant permission. We were simply taking chances. We decided to accomplish our goal through one of the heads of households, Mr. Phiri (not real name), who personally knew the Chief so that he could introduce the research team to the Chief. We needed to act fast because our hotel booking was ending in two days. With the help of Mr. Phiri, we decided to visit the Chief without an appointment. Mr. Phiri phoned the Chief and was told to come to the taxi rank. In the company of Mr. Phiri, we drove to the taxi rank to meet the Chief. When we got to the taxi rank, Mr. Phiri told us that he was unable to accompany us to meet the Chief directly, and simply pointed the Chief to us—"There is the Chief, go and meet him … he is very understanding". We were left helpless because we never rehearsed greeting the Chief using his totem and introducing ourselves in vernacular. None of us spoke Tonga. Nonetheless, we proceeded to meet the Chief. Together with the other female researcher, we hastily put on our headscarves and wrappers as a sign of showing humility and respect to the Chief, and familiarity with culture. Our male counterparts felt in control of the situation. We were simply taking matters for granted! Consequently, there was a confusion when we got to the Chief at the taxi rank because none of us was prepared to directly speak to him. Meanwhile, Mr. Phiri was watching the whole drama from afar. As a team leader, I greeted the Chief and tried to narrate the purpose of our project and that we wanted to interview him. The Chief was livid. He started to rant at us in his local language. We stood speechless, baffled, and looked foolish. He said if we wanted his attention, we should follow protocol, and make an appointment through his secretary. We were really embarrassed. In our minds, we thought the Chief was going to quickly support our request since the project was going to benefit the local community. I immediately realized that protocol, power and authority are more revered above any perceived gains from development projects. I then determined that I will never take traditional leaders for granted in any research context. After the non-event with the

Chief at the taxi rank, the research team carried out focus group discussions with taxi drivers, some commuters and informal traders with the help of Mr. Phiri. We never attempted to interview the Chief again.

I Am Scared to Publish What I Saw During Data Collection

This narrative is based on my academic research experiences. I was appointed as the coordinator for a practical module to undertake research at a secondary school in Limpopo Province of South Africa. My task on creating rapport with the school administrators was made easy because students were simply expanding work on a project with existing Terms of Reference. Because of a well-laid groundwork, we embarked on a successful data collection expedition with the students, and generated hordes of data on the status quo including first-hand photographs. My role as the project coordinator was to bring together themes that emerged from the multiple realities of students' research through a consolidated report. Sharing the photographs through the report was a daunting task because the situation analysis pointed to mismanagement of the school infrastructure. Yet, the school administrators permitted the students to undertake the research at the school under such conditions. Through the consolidated project report, I felt as if I was exposing the ineptitude of the school administrators (because of their inability to run the school) to the higher authorities. In order to distance myself from these realities, I put aside some of the photographs and appended the executive summary by noting that the report was purely an output of an academic exercise.

Presentations of the research results on various platforms (re)shaped the project in many ways. For example, it was suggested during students' presentations that an exit questionnaire be included in the practical exercises in order to extrapolate evidence on the implications for community engagement on teaching and learning—something I had never given any thought at the beginning. This element engendered a paradigm shift to my approach, thinking, strategies, and processes on knowledge transfer through community engaged teaching and learning, as well as on the measurement of learning outcomes from students' experiences. The practical work thus emerged as one of the best practices on community engaged teaching and learning in my Department. As a result, I was given another task to expand the project to capture and reflect on broader discourses that engender sustainable town and gown relationships. In the end, my approach to community engaged teaching and learning will never be the same because of this experience from the practical module.

Don't Rush Us: Seeking Permission to Undertake Research at a Local Municipality

After being awarded a prestigious grant to undertake a research at a local municipality in South Africa, I waited for the disbursement of the project funds for more than six months. The research problematized the spatial expansion of the peri-urban zone into ecologically sensitive areas of a small town in Limpopo Province. The project inception was delayed. Instead of utilizing the waiting period to generate permission to embark on the research project, I procrastinated this activity until the funds were deposited into my research account. I assumed that since a memorandum of agreement exists between the local municipality and our university, obtaining a permit to embark on the project was therefore automatic. However, my initial visit to the municipal offices immediately revealed the bureaucratic procedures I needed to fulfil before commencing any fieldwork. Although cognizant of the red tape to be followed, the municipal officials showed little urgency of the assignment as I was referred to different offices, and went up and down the municipal corridors. I almost gave up. This option was impossible because our international and local project partners were already excited about participating in fieldwork-based data collection. I needed to act fast and obtain signatures before the municipality closed business for a national holiday. It took a lot of patience to wait at individual doors soliciting and pushing for the signatures, which I eventually obtained. The research team took the first day of the data collection programme to familiarize themselves with the research area in physical and conceptual terms. In summary, I discovered that there was a huge difference between the way I conceptualized peri-urbanity in the small rural town and the reality on ground. The research team quickly revised and reconstructed the data collection instruments, and remapped the data collection activities. We eventually discovered that the study area was physically broad, thus creating a considerable challenge to sampling the research participants. It was just a mess. We therefore decided to move in circles focusing more on generating rich research data that could address the research objectives rather than preoccupy ourselves with methods designed to canvass participants for the study. In this regard, the criterion for selecting the research participants shifted to more of a compound of purposive and convenience sampling procedures. The research presented learning opportunities for students and experienced researchers. Our research partners enjoyed the experience of the countryside and the scenic view of the mountains of Limpopo Province. We were also able to disaggregate our findings to generate new themes for future collaborations. It took us a week of data collection and workshopping to make sense of the data obtained from the field. Two students completed their dissertations using the data collected on this project. We presented our research findings at an international workshop. The research findings were well received by the virtual audience.

Conclusion

From my personal encounters, I realized that social research is not practised in a vacuum, but is enabled and enacted within specific social contexts. Undertaking field research is therefore characterized by a homogeneity of experiences that vary in space and time. These can generate intended and unintended outcomes. Each encounter with local communities and stakeholders presents unique, intriguing, and often frustrating experiences. Therefore, field research (no matter the magnitude) requires humility, patience, calculative and intelligent responses, and at times wit. The determinants of successful data collection endeavours are embedded in academic competences of researchers and their willingness to learn from local communities. Regardless of their personal academic competences, social science researchers must not take social and political structural elements for granted.

Correct application of transactive skills by social planners can generate positive reciprocal tendencies from the communities being studied. It is also important to note that local cultures are dynamic and complex. Knowledge and ability to speak the local language (vernacular) of the studied communities are key variables and resources that mediate and influence the success of field activities in significant ways. Language does not only determine the success of fieldwork activities, but also the interpretation and transformation of research results into relevant practical interventions. Apart from enabling rapport and building trust, language also plays a vital role in facilitating communication between the research team and local communities. In addition, recognizing and respecting sources of power and authority is obligatory for institutionalized research results. The role and influence of lead researchers in project teams cannot be underestimated. It is therefore imperative for social researchers to strategically position their structurally informed capacities to work in creative and innovative ways without infringing the research ethics. Each individual research activity must be treated differently. Clearly, social research is a sensitive process and activity. It is neither an art nor a science, but a combination of both.

Emaculate Ingwani is Senior Lecturer in the Department of Urban and Regional Planning in the Faculty of Science, Engineering and Agriculture at the University of Venda. She is a social planner with a Ph.D. in Sociology (Stellenbosch University); MSc in Urban and Regional Planning (University of Zimbabwe); Postgraduate Diploma in Project Planning and Management (University of Zimbabwe); and Bachelor of Education with a major in Environmental Science (University of Zimbabwe). He is a member of the South African Council for Planners (SACPLAN), South African Planning Institute (SAPI); and the Zimbabwe Institute for Rural and Urban Planners (ZIRUP). Her social research competences are continuously nurtured and strengthened through academic and corporate research experiences. In the past, she worked for environmental consultants as a socio-economic research assistant in Zimbabwe, and at universities in Zimbabwe and Ethiopia. Currently, she teaches and supervises research (Honours, Master and PhD levels) in the Department of Urban and Regional Planning at the University of Venda. She also coordinates undergraduate research, and the Integrated Development Project.

Chapter 15
Exhuming Lived Experiential Past in Rural Development Research and Practice

Langtone Maunganidze

Introduction

My experience in rural development which spans more than two decades has oscillated between various roles that have veered towards transformative research and advocacy. These roles availed me the opportunity to interact with various development actors in government and NGOs, and research institutions in Zimbabwe. This was augmented by extensive university teaching and research contacts in Botswana, Zimbabwe and South Africa augmented this. Despite the usually tight schedules associated with such engagements, I occasionally participated in community development projects and managed international research projects. All these often require balancing between one's professional and political interests. My participation at various international conferences has been particularly revealing because it effectively assisted me in thinking in a more enterprising manner, culminating in the publication of many scholarly papers on rural development and entrepreneurship in referenced journal articles and book chapters. The following experiential stories were purposively excavated from my trenches of rural development in Zimbabwe because of their richness and depth as career defining recollections.

The 'Novice' or 'Naïve' Take-Off?

It was in 1996, a year after I graduated from the university with a sociology degree, when I secured employment as a project officer in a Christian civic education and human rights organization. Although I was stationed in Harare, I was expected to make regular visits to rural districts to conduct research and facilitate community education meetings. A development agency based in Europe funded the project

L. Maunganidze (✉)
Midlands State University, P Bag 9055, Gweru, Zimbabwe

T. Madzivhandila et al. (eds.), *Development Practice in Eastern and Southern Africa*,
https://doi.org/10.1007/978-3-030-91131-7_15

which a conglomerate of local civil society groups including churches coordinated. Its implementation co-evolved with the surge in democratization in Southern Africa. Massive voter apathy, especially among the youths, characterised the operational environment that prevailed at that time. I worked under the supervision of a Senior Advocacy Officer. My key responsibilities included data collection, attending community meetings, writing reports and facilitating voter education in rural areas of Mashonaland East and West provinces. Ironically, these were the ruling ZANU PF party's strongholds. The project sought to address the observed absence of youth participation in both national and local government elections in Zimbabwe. Timid opposition in rural areas was evident. Consequently, this created a pseudo one-party state system. The organization's mandate targeted cases of inequality, marginalization and human rights abuses. This invariably extended our foci beyond the church boundaries and consequently required me to interact with multiple development actors, including traditional leaders and party 'foot' soldiers.

One Thursday morning in April, in the company of my mentor, we drove to one of the rural districts in Mashonaland West province and located about 90 km from the capital, Harare. We were on a mission to attend to attend to one case in which there were allegations that the youth in one of the villages were being barred from attending political meetings on voter education especially those organized by NGOs. The journey was filled with a combination of curiosity, anxiety and fear. I had been told and heard through the press about some unconfirmed horrendous stories of similar political intimidation elsewhere in the country. Although my mentor was not a stranger in the village, he confessed that he had not been to this kind of event in the past.

Conveniently and as expected of protocol during those days, our first port of call was the resident or local church pastor's office. The meeting with him was brief and he warned us to proceed to the village with caution as the local 'gatekeepers' especially ruling party activists could be hostile to visitors. However, the 'man of God' assisted with identifying possible informants, who were all part of his congregants and provided us directions to their homesteads. We left the vehicle and any visible research related materials such as cameras at the church. As we arrived at one of the homesteads (unfortunately too close apart to allow any privacy) we thought we had been so luck to meet one of the youths since these were actually our primary target. He was reluctant to help us with directions to our informants' homesteads. Least did we know that all the villages in the area through their local ruling party structures had recently introduced new security measures relating to handling strangers. We however managed to reach out to one of the informants, a church elder. We had a lengthy discussion characterized by repeated expression of discomfort and suspicion. His gatekeeping antics particularly his negative attitude towards our organization was not helpful either. He neither confirmed nor denied the allegations of political intimidation but just strongly advised us to leave. However, he reluctantly allowed us to speak two male youths whom we had found in his homestead. This was done under his watchful eye after which we peacefully departed.

Perhaps our visit could have been either poorly timed or naively elitist. Additionally, we lacked intermediaries to facilitate both access and consent of participants.

These were some of our reflections as we left for the city. On our way back I looked more concerned about our visit than my mentor. We also might not have invested much in appropriate methodological triangulation which is normally effective for accessing 'ring fenced' communities. It was such an enriching initiation into rural development research and practice. I however resigned from the post before end of year to join a government department as an administrative assistant.

Food Aid as a Research Intermediary

Experience in advocacy and civic education in Mashonaland West Province had ably initiated me into rural development research. About three years after our 'near-false take-off' in the first case, I undertook a field work study among rural communities as part of the master's degree. The project took me to another province, southern part of Zimbabwe. During that period, I was also working in a government department based in Harare and would occasionally take official leave to pursue studies. I sought to investigate the dynamics of the grain loan scheme and its effect on rural livelihhods. This was a government sponsored drought mitigation initiative targeting small scale rural farmers and households. Under the scheme, villagers received loans in the form of grain maize and would repay after their future harvests. The programme was coordinated by District Administrators (District Development Coordinators) via local leadership. However, there were numerous stories of beneficiaries defaulting on repayments. This was my first interface among my own people, as a researcher and beneficiary of the programme. I was an ethnographer at home. Consistent with emic perspective, I found out that grain loan scheme was used as a survival strategy by the ruling party, local government structures and rural households.

Numerous methodological issues arose from this trench. The choice of district and selection of the two villages was both purposive and convenient. I selected two villages; my own and the other one which was located about 40 km away but closely situated near the district office. I only selected another village outside my own for purposes of comparison. My own family members and village mates who benefited from the programme became the key research participants. Thus, my field work reflected a flip-flopping between insider–outsider axes. In my own village, direct observation and informal interviews were the dominant data collection methods. As a reflexive researcher, my prior knowledge, experience, beliefs and values coloured my positionality. Although regarded by participants as one of their own and knowing full well that I was 'indigenous', I remained an 'outsider' given my educational background and economic status.

In my own village, access and consent was hardly problematic as villagers regarded me as part of their own voice. Unlike in the other village, where I had to contend with the regular gatekeeping protocols of seeking permission from the District Administrator (DA) (later renamed District Development Coordinator), and the other local level authorities. Luckily, I had once interacted with DA some years back during my tenure as a committee member in a district development committee that had been initiated by a former local legislator (Member of Parliament). Thus, my previous acquaintances with him enhanced my acceptability. Furthermore, the DA was also pursuing degree studies with a local university and this could have made him such receptive but remaining protective regarding the politics around the phenomenon. There were aspects considered a privilege of the educated and elites on one side and on the other, for the ordinary people. Thus, negotiating entry into the district communities was reduced to some 'elite bargain'. Consequently, the inquiry was not free from censorship. Together we went through the research instruments and trimmed some aspects. I occasionally phoned my academic supervisor for guidance and to refine my memo writing and coding. I was introduced to the district office staff particularly those responsible for the scheme and was provided the introduction note to the respective village leadership. I got access to relevant secondary data such as rules and procedures for loan application and repayment schedules. However, this may not have been sufficient to facilitate access and consent at village level. I needed to personally negotiate past the next lower levels of gatekeepers; village heads, household heads or farmers. The village was located just a walking distance from the district office and coincidentally the local chief's village. I first visited the village head's homestead. After some informal discussions he took me to the chief's home, where I met an unassuming and helpful lad. I had not planned to stay much long in this village, but the good rapport extended my stay to two weeks. My stay was punctuated by respect and expectation. The proximity of the village to the district service centre and existence of community development initiatives funded by an international NGO had influenced the villagers' attitude towards outsiders especially donors and researchers. Anticipation of future assistance strengthened our rapport in both villages. Anticipation or expectation of food aid worked as intermediary in negotiation access and cooperation.

My interaction with villagers taught me that community members were receptive to researchers when they hoped that these would facilitate transformation of their livelihoods. I completed my data collection in both villages within two months and finalised my draft report. Respondent validation or member-checking was arranged with a few purposively selected community leaders and beneficiaries from the second village since the first one represented a control group.

Can Rural Development Research Be Value-Free?

This project began as a response to an international call by the Organisation for Social Science Research in Eastern and Southern Africa (OSSREA) in June 2011 for contributions to a book project on a theme: State fragility in Eastern and Southern Africa. I received the acceptance of my proposal and abstract with great excitement and expectation. The project was timely in the light of unprecedented economic and political crises that obtained in Zimbabwe between 2000 and 2010. I sought to track events in the country since the turn of the new millennium examining the drivers, dimensions and effects of state fragility. I endeavoured to do an inventory of discourses constituting the intersection between institutionalized corruption and the perpetuation of state fragility in the country. I combined documentary survey, direct observation through site visits and interviews in selected villages in two selected provinces adjacent to the capital Harare.

At the start of the project I was in the diaspora, in Botswana, teaching at a university and only occasionally visiting Zimbabwe mainly for purposes of either conducting rural development research or as visiting scholar. My entry into the field, choice of problem area, research questions, and sampling were value laden. I entered the field with some preconceived ideas which influenced the proposed running title; State institutions as harvesting roads. This also could have possibly led to a seemingly and unconsciously hurried data collection process. The timelines for the project did not allow pilot study and preliminary data collection. This was my first time to engage in reflexive auto-ethnography, relating my experience and critically investigating the discourses that have constituted that experience. While documentary evidence was easily accessible the main challenge was conducting interviews especially with state and political elites (Mikecz, 2012). This was a sensitive subject given the level of political polarization and censorship that had engulfed the country since turn of millennium. However, it was not difficult to access villagers and local leaderships. My familiarity with the population in the general and particularly past research contacts as an 'insider' facilitated access and availability of respondents. My acquaintances tried though in vain to serve as intermediaries for my access to political elites. It was difficult to avoid using covert observations and discussions. Furthermore, my research assistant was also struggling with the task and meeting the deadlines. I needed a full paper before end of March, following some member checking and presentation of findings to colleagues ahead of the conference pencilled for mid-April in Addis Abba, Ethiopia. The flight to Addis a week later was filled with anxiety and excitement. In a funded project one would not only wish to do just a good job but also meet the expectations of the donor. The presentation was well received.

The anticipated period of publication changed from months to years albeit frustrating back and forth manuscript revisions as the editor and reviewers continued to demand more empirical evidence and reorganisation of the material including the change of title to; Zimbabwe: institutionalized corruption and state fragility. The publication finally came out in December 2016.

Lessons from this trench are varied. Documentary evidence especially from official and media sources tended to be partisan especially during election times. Rural development research is complicated by numerous layers of gatekeeping. Ruling party officials acted as both key informants and gatekeepers and, in the process sought to capture or manipulate the sampling and data collection processes.

Moral Compass or Blank Signpost?

This experiential story marks my second return to my home village. It was a self-funded reflexive ethnography targeting three purposively selected villages. I sought to explore the extent to which most of the indigenous knowledge systems (IKS) in rural areas in Zimbabwe had been utilized or discarded as a guide for addressing people's daily challenges such as health. My engagement was anchored on the following development question; whether IKS still guided the development path of rural communities? I selected two villages from the southern part of Zimbabwe and for two main reasons. Firstly, the researcher's familiarity with the target population and previous research contacts facilitated accessibility and availability of respondents. Secondly, key informants were chosen based on their willingness to provide the information by virtue of knowledge or experience. The greater part of my boyhood had been spent in the villages with only occasional absences from the village while a student at the nearby boarding secondary school. We had grown up using indigenous wild plants either as food or medicine.

My adult life experience as a rural development sociologist motivated me to begin reflecting on the phenomenon. I can easily conclude that I had been involved in experiential learning through informal investigation on the subject for decades. Interest in conducting systematic research on the subject only struck me around 2014. I had just attended a workshop on IKS and Public Policy organised by the University of Botswana (where I was then teaching Rural Sociology). I was also taking a graduate class on Development Policy and Practice and wished to enhance my teaching with a research and publication output on the subject particularly exploring the factors influencing the marginalisation and exclusion of IKS from mainstream development process. My study was guided by the social constructivist paradigm and followed a reflexive methodology, which recognize the complementarity of locals' observations, categories, explanations, and interpretations and those of the researcher. This provided for a deep understanding of the people studied. This was perhaps the first time I began to fully embrace this kind of methodology. But is it also not a useful guide for rural development research?

I planned to collect my data during the university vacations; June-July 2014 and January 2015. Qualitative data were collected using informal interviews and individuals' lived experiences. I purposively selected individuals or groups of individuals that I believed were proficient and well-informed about the subject. They included village health officers and nurses at the local clinic. Informal focus group discussion participants were selected using the convenient sampling technique. They were all familiar with me and aware of my agenda. With this technique participants were selected simply as they just happen to be situated near to where I would be during the time of collecting data. I utilized events such as funerals or village meetings. Discussions among men at a funeral gathered around a night fire were not formally structured but produced credible and quality insights on issues. Men at these events engaged in serious debates on subjects ranging from national and local politics, religious and cultural, and sport to kill time. However, this consequently often excluded the women's perspectives. I attempted to fill in this gap by interviewing key female informants particularly on issues relating to reproductive health and childbirth practices. Participants' perceptions and explanations varied by age. For instance, the older generations were able to explain the loss of IK and attributing contributing factors to external factors while the young generation criticised the older generation's failure to transfer information. Key informants on traditional healing and medicine were drawn from the older generation. I also reflected on my previous research in the area and a recollection of "lived" experiences.

I have learnt that in reflexive research, positioning and standpoint (Hennum, 2014) were central elements to framing relations between researchers and informants. Through informal discussions with one's own people generated credible qualitative data. However, researching or practising at 'home' also generated biases and provided slanted data since historical association with the researcher could lead one to be misconstrued as an advocate for development. Failure to fulfil this expectation may lead to lack of confidence which could jeopardise future engagement and cooperation.

Mid-Wives of Rural Peripheralization

My previous participation in book projects and the consequent recommendations for further research motivated me to expand the scope of my future inquiries. Thus, I began investigating on the role of political elites in rural development drawing threads from the previous studies. I also purposively and conveniently selected a district situated in the Masvingo province, southern part of Zimbabwe. I sought to examine the encounter between state-sponsored elites and the peripheralization of rural communities in the district. State 'sponsored' elites have both direct and indirect backing of the state and often operate within the structures and institutions of state. This was not just done to gain knowledge but also for purposes of publishing. I did not receive any funding for this study.

While the other previous investigations were more structuralist focusing on institutions, this one was done through the lens of a Norman Long's (1992; 2001) actor-oriented interface approach, a direct departure from the traditional or classical power elite and state centric theories (Kifordu, 2011). My observation was that rural development programmes in the district failed because they were driven more by elite interests than 'community' ones. While these elites have tended to project themselves as assets or catalysts for rural development, they have generally accentuated the peripheralization of ruralites. However, the 'peripheries' are not necessarily passive actants. Locals have the tendency to feign loyalty. There have been numerous cases of political elites or ruling party 'heavy weights' losing local elections to very lowly placed locals after underestimating their potential to resist. Notable attempts to resist the tendencies of such state-centric elitism failed as local representatives became culpable in influencing the practices and materials of this peripheralization. Political elites have therefore become mid-wives of rural peripheralization. State 'sponsored' elites effectively reduced locals to permanent 'hinterland' or 'dependent' stakeholders (Maunganidze, 2020). Elites have continued to act as intermediaries, gatekeepers and brokers to accessing rural communities. Similar to my experience in the first trench, failure to negotiate entry and access to the vulnerable was attributed to my positionality with respect to local elites.

This experience taught me good and bad lessons. Firstly, relying on documentary survey poses validity challenges. Institutions and individuals have tendency to positively potray themselves leading to slanted data. However, I also learnt that in rural development, documentary survey can be as authentic and original as primary data. Secondly, although I did not need to get past the hurdle of official gatekeepers, I could have got trapped into becoming my own gatekeeper; or a state of 'self-blocking'. Thirdly, solely relying on data from my own previous studies could discourage critical reflection on my own social and political commitment and preconceptions (Saukko, 2003). Fourthly, and on a positive note, the use of theoretical framework such as the actor-oriented interface approach provided a systematic conceptual and methodological framework for deciphering and analyzing the interlocking of life-worlds and actors' 'projects' from documentary evidence.

Conclusions

The above cases reflect the challenges and opportunities that shaped my journey from a 'novice or naïve' field officer to becoming an established researcher in rural development. Regarding rural development research, these experiential stories represent a mixture of methodological complementarities and supplementarities. The power of ethnography is particularly evident. At the time of my interface with communities I hardly anticipated that i would one day in future be expected to 'exhume' these experiences. The adopted narrative perspective would hopefully illuminate future researches. The metaphorical sense of this piece as reflected by the title could act as moral compass for future engagement in rural development research and practice.

This testimony is not a claim to authenticity and truth but just self-reflexive introspection. However, the stories may fail to distinguish between 'lived' experiences and texts of my own creation as I navigated between the shoals of emic and etic perspectives. Given the contiguities and overlaps between rural development research and practice, there is potential to confound or conflate them resulting in one missing out on the inherent blind spots. My own competence and social values influenced my ontological and epistemological positions and consequently the generation of research ideas and topics. It is hoped this would be mitigated by the complementing and supplementing stories of colleagues with different training and experience in development practice.

References

Hennum, N. (2014). The aporias of reflexivity: Standpoint, position, and non-normative childhoods. *Journal of Progressive Human Services, 25*(1), 1–17.

Kifordu, H. A. (2011). *Nigeria's political executive elite: Paradoxes and continuities*. Erasmus University.

Long, N. (2001). *Development sociology: Actor perspectives*. Routledge.
Long, N., & Long, A. (Eds.). (1992). *Battlefields of knowledge: The Interlocking of theory and practice in social research and development*. Routledge.
Maunganidze, L. (2020). State 'sponsored' elitism and rural peripheralization in Zimbabawe. *The Dyke, 12*(1), 97–116.
Mikecz, R. (2012). Interviewing elites: Addressing methodological issues. *Qualitative Inquiry, 18*(6): 887–910.
Saukko, P. (2003). *Doing research in cultural studies*. SAGE Publications Ltd.

Langtone Maunganidze is a Senior Lecturer and Program Coordinator at the Midlands State University in Zimbabwe. He holds a PhD in Sociology and Masters in Sociology and Social Anthropology degrees, obtained from the University of Zimbabwe. His teaching and research interest areas cover issues in industrial sociology, rural development and entrepreneurship. His service to academia and profession has taken him to various countries as a visiting scholar, external examiner for advanced degrees and reviewer for internationally acclaimed publishing houses. He has ably mobilized and managed internationally funded research projects and student exchange scholarships. He is also a guest researcher at the Global Centre of Spatial Methods for Urban Sustainability (GCSMUS) at Technische University (TU), Berlin, Germany.

Chapter 16
Centrality of Experience in Shaping Rural Development Praxis

Ephraim Chifamba

Introduction

I have been in the field of rural development for more than 18 years. After graduating from the university, I worked for five years as a community development officer and was responsible for giving relief to children living under difficult circumstances. The same project targeted minors living in streets and those affected and infected with HIV and AIDS. Further, I conducted case work analysis, which facilitated the screening of beneficiaries.

After years of unique exposure as a result of working with vulnerable children, the drive and commitment for offering assistance to vulnerable communities continued to compel me to seek new challenges. The experiences acquired from an exciting career working for CARE International, which lasted for more than 10 years, further presented opportunities to work in people-oriented programs with various stakeholders, enabling me to traverse the whole of Zimbabwe. This afforded me the opportunity to work in multi-cultural setups. The vast experience obtained during my humanitarian work helped me to have answers to some of the research and development questions that revolve around rural gender dynamics, land tenure and rural research methodologies among many other issues. The determination to expand the horizons of knowledge through research, teaching and community service further pushed me to seek greener pastures. I got a job as a lecturer at one of the state universities in Zimbabwe. The time spent in the academia enabled me to engage in research work that culminated in the publication of several papers in refereed journals. The "lived" experience in both practice and theory places me at a vantage point to share and reflect on my endeavour in rural development work. The lessons learnt should serve as valuable reference tools for policy makers, professionals and researchers in

E. Chifamba (✉)
Department of Rural and Urban Planning, Great Zimbabwe University, P.O Box 1235, Masvingo, Zimbabwe

T. Madzivhandila et al. (eds.), *Development Practice in Eastern and Southern Africa*,
https://doi.org/10.1007/978-3-030-91131-7_16

the field of rural development. The following stories provide perspectives emanating from primary field researchand community engagement.

Navigating Confidentiality and Ethical Consideration in Development Research

As a point of departure, one might immediately ask the following questions: How should a social science researcher react when he/she cracks a puzzle in a murder case, which is still undergoing police investigation? Does reporting the murderer to the police violates confidentiality? I was drafted into the Masvingo Rapid Response Team (MRRT) as part of the Government of Zimbabwe (GoZ) initiative to assess the effects of COVID-19 pandenic on the livelihoods of people living with disability. My team was tasked to collect data in Chivi and Bikita Districts. These are the poorest districts in Masvingo province. Precaution measures were put in place before departure to avoid contracting the COVID-19 pandemic. WhatsApp platform, desktop and in-depth interviews were used for data collection to avoid health risks and abide by government stipulated lockdown restrictions. The survey was conducted after realizing that COVID-19 pandemic impacted negatively on human rights, human lives and livelihoods of people living with disability. These people struggled to access health services, food and education among many other requirements during the COVID-19 lockdown. The survey intended to provide critical data in order to inform government decision making process.

We set out to collect data in the two districts in the early hours of 28th May 2020. One of the respondents was a 58-year old woman. She was handicapped five years ago when her cancer wreaked leg was amputated. She relied on a wheelchair to move around. Like many other poor households in the two districts, her situation was not different. The woman mourned that the spate of COVID-19 prevalence heightened the despair in her household. The woman further pointed out that the failure to access relevant information on COVID-19 due to poverty, isolation and illness imperiled her existence. Instead, she depended on information from neighbours as she did not have a radio, smart phone or a television set.

The old woman avowed that she subsisted on vending fruits and airtime at the nearby business center before the occurrence of COVID-19. However, the closure of businesses and social distancing regulated people's movement, thereby affecting her daily earnings. Her disability had affected the prospects of accessing existing coping mechanisms as was the case with other able-bodied females in the district. Loss of reliable earnings compromised her chances of raising the USD $200.00 that was required for fixing her simulated leg at Driefontein Mission Hospital. She further complained that the lockdown barred her from sourcing funds for the treatment of her elder son who was born with osteomyelitis (a condition in which the right leg of a person is shorter than the left one). Hunger was also singled out as another challenge, which affected her. She further restated that the COVID-19 pandemic lockdown was a serious threat to women with infirmities ahead of all other social groups including senile men experiencing enervations. She reiterated that other able-bodied women were involved in 'nocturnal vending' but she could not do the same due to her ailment. Other hawkers could escape the harassment from state agents, a tactic which was difficult for her to embrace due to her illness. The respondent repeated that the outbreak of COVID-19 pandemic compelled women with debility to experience 'four lockdowns' that is, womanhood, disability, segregation and the actual lockdown.

After the interview, we packed our gadgets in preparation to visit the next respondent. It was at this moment that one of our group members requested the whereabouts of her husband. With tears streaming down her cheeks, the respondent revealed that her husband was an accused person at Mutimukuru prison. We further requested for the reasons behind his incarceration. She recounted a bloodcurdling story. Her husband gave his nephew three cows to keep in trust. Unfortunately, the nephew passed on the following year. Out of compassion, her husband allowed the remaining widow to use the cattle for some years. He decided to collect the cows five years later after the passing on of his nephew. The nephew's widow rebuffed to surrender the livestock on the understanding that she was not privy to the agreement. Her husband sought succor from neighbours and pressured the helpless widow to surrender the cows. The widow then reported the matter to the police and a stock theft case was raised against her husband. The investigations revealed that upon receiving the cows, the nephew registered them in his name. The matter was tried and her husband remanded pending his being imprisoned. Of course, this matter was not part of the information that we intended to collect from the respondent. However, we thought that justice did not prevail in this case. We referred the respondent to Lawyers for Human Rights for assistance in handling the case. The organization concurred that justice did not prevail and promised to represent the accused after the end of the COVID-19 pandemic lockdown period.

The Implications for practice is that this case raises some social science research questions. What should researchers do if they learn about illegal activities during an interview session? In the process of interviews, researchers come across matters of public concern which require redress or justice. Reporting the criminal to the police may violate confidentiality, which is one of the key ethics in research. However, social science code of ethics allows for a breach of privacy and confidentiality, for example in the case of rape. Furthermore, the law obliges researchers to reveal information about illegalities, in certain circumstances, to give law enforcement agents research material. As things stand, precincts on privacy and confidentiality constitutes a prickly subject among researchers.

Like Hens, Women Wait for Cocks to Crow Announcing the Arrival of Daylight

In 2007, I was part of a four-member research team that carried out an exploratory study on the impact of drought in Matabeleland North Province of Zimbabwe. The study covered Binga, Nkayi, Tsholotsho and Umguza. The study was carried out as part of the Zimbabwe Vulnerability Assessment (ZVA) survey, which was a program led by the government and its partners. The study targeted hard-to-reach and under-served households in the four districts. The research team and facilitators were enlisted and trained to lead in-depth interviews, focus group discussions and provide initial analysis of the study results. The community entry procedures were carried out smoothly before conducting the research.

We were scheduled to visit five households on the first day of the interviews. We left Bulawayo early in the morning and drove 220 km to Tsholotsho District. We arrived round 10 am in the morning and started identifying the respondents. As part of our community entry, we had already identified the households for interviews. Thus, it was easier to contact the targeted households. We agreed that each member of the team was supposed to interview one household. I opted for a household that was adjacent to the main, dusty road.

A woman quickly came out to greet me as soon as I arrived at the targeted household. She opted to use her thatched kitchen as the venue of the interview. I accepted politely and got into the kitchen. She hastened to inform me that her husband would join the interview after watering few plants in their garden. Without wasting time, I requested permission to start the interviews. In the middle of the interview, her husband arrived, and I greeted him. However, he came across as being offish. I quickly discerned that all was not well. He requested my identification even though we had once met during our community entry engagements with the participants. While standing, he asked why I was interviewing his wife in his absence. I quickly apologized using the Shona language even though I was right inside a core Ndebele community. I indicated that I took it for granted that he was aware that I was in the area as part of Government-NGO initiated surveys. I further reminded him that Chief Mathuphula had accepted that the interviews should proceed in the area. However, he nodded his head, showing signs of disapproval and derision.

The man started advancing forebodingly towards me, holding a walking stick in his hand. In the process, he started yelling and calling me all bad names as ascribed by the Shona tribe. I quickly realized that I had broken socially ascribed roles of men and women in the community. I realized that answering back could invite more trouble. I picked the writing pads and a voice recorder and left the household. After running for about twenty (20) metres, I realized that the camera was left in the kitchen. I ran towards the car, which was parked by the roadside and the driver drove off towards Tsholotsho growth point. We arrived at the police camp and reported the case. The policeman in charge instructed one officer to escort us to the household to recover our camera. Upon interrogation, the man indicated that I was not supposed to interview his wife in his absence. He pointed out that:

> My wife can do all the work, except entertaining outsiders in my presence. Values in the community do not permit a woman to engage male outsiders, just as the same values do not allow men to wash dishes, cook and sweep.

In its purest form, this statement points to gender inequality and culture existing within the communities. After collecting our camera, we left the homestead. The research of this nature requires the observance of the values and norms of a community rather than perceived from an outsider's viewpoint.

Indeed, this experiential reality raises the question of 'out sidedness' in community research work and it prepares researchers for problems they should anticipate facing as visitors. Rural development practitioners should be conscious that, as 'interlopers', they must first witness some discomfort as they enter a community. Such uneasiness is normal and not an experience to be escaped. It is a necessary and normal phase in building a rapport with the community. Conceivably, data collection training programs should deliberate on the numerous approaches that are useful in ensuring community trust and assess their potential effects on stakeholders. If rural development practitioners accept local norms and values, do they conform to the accepted communal procedures and remain faithful to the research ethics? What are the ethical effects of conforming to communal values and norms? How do these values and norms contrast with the role that should be played by the researcher? These are fundamental questions that seek answers.

An Interviewee Crying from Hunger in a Room is Hard to Shut Out; It Pins Responsibility and Skill on Interviewers to Collect Reliable Data

In 2015, Musasa Project financed a study on gender violence in Masvingo province, covering 3 districts. I was a member of the technical team. We travelled to collect data in one of the targeted districts on the 15th of August of that year. The first respondent was a 67-year old disabled widow with one adult son and two grandsons. She was aware that we were coming to interview her.

Members were given the chance to introduce themselves and explain the reason why they were in Chivi District. After the introductions, one member of the team requested the interviewee and other people who were around to introduce themselves. However, she started narrating her ordeal. She recounted that she could not walk because of poliomyelitis that developed at the age of three. Her frailty worsened as she was growing. Other infirmities such as heart failure, meningitis and HIV and AIDS further aggravated her condition. The woman further narrated that surviving in rural environments with these illnesses posed severe threats to her life. She was failing to access health facilities, proper education for her grandsons and food. She said that Women's Coalition of Zimbabwe (WCZ) used to assist her with foodstuffs. However, the economic conditions forced some donors to roll back. "Since the beginning of the year", she said, "no organization made an effort to bring people living with disability (PLWD) together for them to share their challenges". The woman went on narrating how she used to earn between USD $1.00 and USD $3.00 for attending WCZ workshops and meeting. People living with disability were no longer called to these meeting due to donor fatigue and limited resources. To her, the absence of these meetings affected another source of modest income. Lastly, she complained that hunger was also affecting her since she survived through handouts. She requested her son to show us a 5-kg packet of maize meal that was almost empty.

She spent almost two hours narrating her afflictions whilst we listened assiduously. Considering the woman's health condition and the food insecurity status at the household, members of our research team ended up sympathizing more with her. Instead of proceeding with the interview as scheduled, we contributed few dollars that were part of our subsistence. We eventually left the woman without interviewing her.

The narration of the ordeal that the interviewee went through ended up influencing the direction of the interview. What does this experience imply for practice? Do researchers need to detach from emotions and do their job? Or do they need to empathise with their respondents and then overlook things that are also essential in their conduct?

Data Collection Dishonesty is not About Good Versus Bad People; It is About Conflict of Interest and Motivated Reasoning

Fraudulent research practices are common among both researchers and enumerators. Dishonest actions are not exclusive to big research projects. Unethical behaviors are all too prevalent because enumerators encounter situations in which they opt to have monetary gains by cutting corners. The temptation for research dishonesty is prevalent, especially in funded researches.

In 2004, Musasa Project conducted a study on the livelihoods of single headed households in the Midlands Province of Zimbabwe. I was requested to lead one of the teams that gathered data in Gweru and Zvishavane. We enlisted four third year students who were on industrial attachment with Musasa Project. The research team was trained on both qualitative and quantitative methodologies and all preliminary research prerequisites were met before commencing the study.

The dates for the interviews were set. We left Gweru on the 13 of March 2004 to Zvishavane, which is a mining town in the heart of the Midlands Province. Each member in the team was given an allocation of funds that were supposed to be paid to respondents after each interview. Respondents were supposed to be given US$10.00 after the interview. Nonetheless, we faced a few challenges in locating the respondents. While good roads criss-crossed the mining compounds, the homes of respondents tended to be over 200 km from the major road. We made arrangement to meet with respondents in their homes. Despite the determination made to get correct directions to the homes of respondents, we learnt that the directions given were imprecise and arduous to follow. Numerous respondents who we spoke to through the phone were unable to give precise directions. We wasted a lot of valuable time trying to locate participants. We had initially targeted five participants per day, but the number eventually came down to less than three participants. However, we were surprised that all university students surpassed the target every day. On the average, they would interview between seven (7) and ten (10) participants per day. On the last day, one of the students revealed that two students fought the previous night. We further inquired for the reasons why a fight broke out during the night. One of the students who also claimed to be a victim exposed that all students connived to fill in the questionnaires during the night. They disguised and pretended to be collecting data during the day. It was later realized that the students deliberately falsified and tampered with data. They converted the money that was due to participants for their personal use. It came to light that the students were surpassing the target in order to raise more money. The study was cancelled, and the students were requested to reimburse the organization. The students were further suspended from the industrial work-related learning for twelve months.

The implication for practice for this experiential story shows that unethical practices are is a commonplace among researchers and enumerators. Unethical practices in data collection are a stumbling block in genuine research work. Although this story might not be representative of research ethics among enumerators, it provides a glimpse of challenges affecting research. It is, therefore, necessary to educate enumerators about the ills and consequences of data collection dishonesty before the commencement of any fieldwork.

Researchers' Views of the Poor as Indolent, Fatalistic, Spendthrift and Answerable for Their Poverty Are Incorrect

After completing my first degree, I worked for CARE International as a data collection officer. My responsibilities involved collecting data on specific studies that were conducted to inform both policy and practice within the organization. My work involved a lot of travelling to rural areas. Having grown up in an urban setting, I had certain perceptions about the rural poor. I saw life differently from the point of view of the rural poor. I viewed the poor as inconspicuous, inarticulate and unorganized. This standpoints also affect other rural development practitioners who are neither poor nor rural. This practical story narrates the experience I encountered with one of the rural poor households that ultimately transformed my perception.

In June 2008, CARE International carried out vulnerability assessment surveys throughout the ten provinces in Zimbabwe. This was a lean period that was characterized by hyper-inflation and shortage of basic commodities. The surveys were structured into different phases that included data collection and analysis, intervention (identification of needs and opportunities) and impact assessment. I was selected to participate in the first phase. We carried out all prerequisite community entry procedures before commencing data collection.

One morning, we left Gweru to Gokwe North Province to collect data on household vulnerability. My first interviewee was a 68-year old widow who suffered from severe breast cancer. Her household also suffered multiple deprivation traps. I requested the woman to narrate the challenges that were affecting her household. She stated that she perceived the the happenings that were misfortunes to others were calamities in her own case. Her husband and son were involved in an accident in 2000 and the husband died on the spot while the son suffered multiple fractures. The woman stated that her challenges started with the passing on of her husband. His son stayed in the hospital at a cost of US$100 per week, which was an amount almost equivalent to their annual profit from agricultural activities She sold household property and cattle to meet hospital bills. The cattle were undersold, but under normal circumstances could have been sold for more than twice the amount received by the household. Parting with cattle meant losing their last capital investment in agriculture. She further recounted that the prospects of buying cattle were unconceivable with the current market value estimated to be over US $400.00. Her son spent over six months in the hospital and prolonged stay worsened poverty ratchets in the household. Disgusted and angry, the woman later took her son back to the village where he was treated with indigenous medical knowledge. The boy suffered permanent physical disability. He could no longer lead a productive life, although he would continue to eat as long as he lived. The woman laid bare all the challenges that affected her. I also met a few similar cases of households that were in the deprivation trap. The experience changed the way I perceived the rural poor.

Rural development profession suffers from deep-rooted preeminence complex, which has a real implication for practice. Most researchers and development practitioners are often in similar difficulty. Rural household face a myriad of challenges, which compel them to disposes pecuniary assets. This leads households into a cycle of poverty. Rural poverty should be viewed as deprivation emanating from both internal and external factors. Researchers usually conduct studies and intervention efforts as if they know everything and their clients know nothing. To get beyond stereotypes and counter-stereotypes, development practitioners need to conduct situational assessment of factors that lead households into deprivation trap.

Conclusion

This chapter highlights lessons learnt during data collection in different studies. The chapter shows that research in rural areas presents distinctive and diverse challenges for development practitioners. The experiential stories point to the intricacies associated with rural development practice. Variations in cultures across the wide spectrum of rural settings, utility of tools developed by urban domiciled researchers, identification of reliable enumerators and ethical issues related to conducting research are some of the problems affecting rural work. In order to uphold high scientific standards and become real change agents, it is vital to respond to field work problems timeously, with locally germane and innovative solutions. Seeking community by-in is indispensable when working in societies unaccustomed to research. These experiential stories provide valuable information for other researchers and rural development practitioners who plan to carry out research in similar contexts. As more is reflected about the convolutions of rural development research and mediations, it is important that these lived experiences be shared and used to conduct reliable rural research and sustainable interventions in the future.

Ephraim Chifamba is a lecturer at Great Zimbabwe University. Ephraim is a recognised scholar with a strong transdisciplinary understanding of developmental challenges and sectors, especially on the African continent. He has more than 18 years experiences in water resource, gender and governance issues in the region. He is author of a number of authoritative publications, including several journal articles, books and book chapters, and has supervised and co-supervised numerous Masters and Ph.D. students.

Part IV
Navigating Social and Political Dimensions

Chapter 17
Agriculture Research, Extension, Policy and Development Practitioners

Tshilidzi Madzivhandila

Introduction

The lived experiences, which I share in this chapter span more than two decades of my professional career in the field of Agriculture, Research and Extension for Rural Development (AR&E4RD). My career advancement has been shaped by great mentors, a supportive team and hard work to mention a few. This path has also required many personal compromises including reduced income and being away from family for long periods of time. For an early career person, these are valued must haves, especially in terms of the economic benefits derived and stability gained from such disposable incomes. My career progression has been moulded by active involvement in the design, implementation, monitoring and evaluation of AR&E4RD projects and programmes. Through these years, I have witnessed successful and failed projects. I am a firm believer in the notion that capacity of individuals, systems and institutions is core to the successful implementation of development programmes.

Professional Development Programmes

After graduating with a Bachelor of Agriculture Management Degree and a Higher Education Diploma, in 1995, I was destined to be a teacher as it was the default career for many young graduates. At the time, my three elder siblings were already teachers. After teaching for two and half years, my career trajectory changed after a casual visit to a close friend who worked for one of South Africa's research councils. During this visit, I leant of a newly launched Professional Development Programme (PDP) that was hosted by the Agriculture Research Council, and which was enrolling the first

T. Madzivhandila (✉)
141 Cresswell Street, Weavind Park 0184, Pretoria, South Africa
e-mail: tmadzivhandila@fanrpan.org

T. Madzivhandila et al. (eds.), *Development Practice in Eastern and Southern Africa*,
https://doi.org/10.1007/978-3-030-91131-7_17

cohort of early career researchers. The aim of the programme was to provide opportunities to suitable young candidates from various disadvantaged backgrounds to develop skills, potential and talent to address the structural effects of apartheid in the agriculture sector. The programme was designed to address the scarcity of adequate qualified and skilled natural scientists and agricultural economists in the agriculture labour market. I was successful in my application and selected to participate in the programme as one of the second cohort of students.

As evident from the many graduates actively participating in leadership roles across the globe, this PDP has successfully achieved its objective. However, despite the well-known benefits of these programmes, several PDPs are not able to achieve any impact at scale. A common weakness among some of these programmes is the lack of support systems after the internship duration. For instance, it is a commonplace that a graduate goes back to the unemployment pool after completing an internship.

Based on my experience, the potential for a successful PDP include: appointment of applicants on a full/fixed term contract basis for the duration of their study programme; the participants should be mentored by seasoned researchers or practitioners; candidates should be evaluated throughout the programme; the training programme should address nexus issues and should integrate personal growth elements; and placement on permanent employment as junior researcher or practitioner where possible is imperative.

Rural Development Participatory Approaches—What Have I Learned?

If well executed, participatory approaches are important to successful implementation of sustainable development programmes as they bring collective decision-making process and ownership of intervention by engaging stakeholders (including community members, researchers, and others) during the design, research and implementation process of a project. When I started my career in AR&E4RD, in 1998, community-based participatory approaches were at the core of development practice and community capacity building initiatives. These initiatives, among others were central to supporting smallholder agriculture systems.

There are numerous participatory approaches that as rural development practitioners can learn from. Early in my career, I deliberately invested time to learn and equip myself with tools and skills for using participatory approaches in the AR&E4RD sector. For instance, I remember vividly my participation in, (i) the pilot Participatory Extension Approach (PEA) using the '*BASED*' approach that was being championed by GIZ (German International Cooperation); and (ii) the Agricultural Research for Development (ARD) programme supported by the International Center for Development-Oriented Research in Agriculture (ICRA). I also participated in two phases of a developmental project called "Beef Profit Partnerships

(BPP)" implemented by a national research institution and funded by the Australian Centre for International Agricultural Research (ACIAR). Core to the BPP project was the participatory capacity building approach called Continuous Improvement and Innovation (CI&I) composed of distinct but related steps as detailed below.

The six steps of the CI&I process

Using CI&I there is an opportunity to make a real difference if attention is paid to the following stages of project implementation:

1. *Situation Analysis*: Need to understand current practices, processes, systems and performance, while identifying the opportunities for action to improve the situation.
2. *Impact Analysis*: Need to prioritize—using robust criteria and evidence—the opportunities with high benefit/cost (i.e., pay-off) to invest in.
3. *Action Design*: For each selected opportunity, there is a need to design specific actions to implement that will make a real difference to improving wealth, wellbeing and the environment. Inherent here is also worrying about how the effects of designed actions will be tracked and measured.
4. *Action Implementation*: This is where the tyre hits the tarmac; i.e., selected actions are implemented while measuring if they are making any difference. Important is to stop doing those things that don't make a real difference.
5. *Results Assessment*: The results of the implemented actions are measured quantitatively and qualitatively. What is often ignored is to understand the drivers of the observed change.
6. *Creation and Synthesis*: To ensure long-lasting change, new questions to answer and ideas should be generated. There should be a structured way to establish new needs and opportunities to focus on next; hence continuous improvement and innovation.

Some of the question that I have come across are: what are the valuable lessons that one can pick from these types of initiatives? How can we mainstream the core principles of participatory approaches sustainably beyond specific funded initiatives, and what are some of the barriers and opportunities? The sad fact is that in most instances, the application and use of these valuable approaches is limited only to the project(s) funded activities/timelines and they do not always apply beyond the life of the project, which is critical for sustainability. In addition, there is still a fundamental failure to mainstream and scale-up these initiatives in core national institutions/programmes. There is a need to systematically and continuously evaluate these participatory interventions across countries, sectors and cultures in order to share lessons and enhance the body of knowledge on these practices. By reflecting on my career progression and experiences, I will share some lessons I have learnt during this experimental part of my journey that could be of value to the development practice.

Participatory approaches (ARD, PEA and CI&I) provided me (and other practitioners) with skills for using a systems' approach towards understanding AR&E4RD. Some personal valuable lessons I have lent from these approaches include:

- *Capacity Building*: I managed to remain committed and motivated because I was equipped with thinking skills, processes, techniques and tools to achieve on-going improvement as an individual, in teams, partnerships and networks.
- *Internalisation:* I have personally used the approaches, moving from the known-to-unknown pedagogy, for both my Master and PhD degrees. Further, when designing, implementing, and monitoring and evaluating AR&E4RD, I use the participatory approaches and gender lenses, to this day.
- *Community of Practice (CoP)*: I have remained within the rural development community of practice where I co-publish, co-supervise and examine academia work in this field. In addition, I precipitate in projects and programmes as part of an advisory team, using participatory approaches. CoP is important for sharing knowledge and ideas and for utilising specialist knowledge, skills and support.

Enabling people (i.e., through capacity building) to develop a clear mission and focus is the most critical success factor in achieving developmental outcomes. This should be accompanied by having a clear and shared understanding of a desired outcome and the boundaries around it. Involving the appropriate people in the most appropriate way is critical to the success of continuous improvement and innovation programmes. Further, equipping individuals in teams with the knowledge, skills, process, roles and responsibilities to practice participatory development can dramatically increase the efficacy of AR&E4RD programs.

How I Learned to Influence Policy

Policy influencing and co-influencing, other than policy research and analysis, is hardly part of an academic curriculum. Further, in my early and mid-career years, I did notice with dismay how research was an end in itself but not a means to an end. An end here referring to societal betterment and the deficient nature of research to influence policy and decision making, also referred to as "research translation".

After I joined an organisation (the Food, Agriculture and Natural Resources Policy Analysis Network (FANRPAN), in February 2011, whose mandate was to conduct policy research and promote advocacy at the continental scale, I attended the first meeting, where my then CEO kept emphasizing that our work without fail should co-influence policy making. How could I have asked what this was amongst other staff members? The first thing I did when I got home was to search for its meaning—and the Oxford Dictionary—which came to my rescue. Two terminologies which will remain in my vocabulary forever are:

- *Co- (as a prefix) which means "together, jointly or mutually; and indicating partnership or equality".*
- Influence to mean *"the capacity or power of persons or things to be a compelling force on or produce effects on the actions, behaviour and opinions of others".*

One of the meeting I had with the CEO, exposed how disconnected I was from the reality of bridging the gap between science and policy. I realised quickly that I needed to avoid the tragedy of the PI (principal investigator) mentality. As part of my discussion with her, I asked the question: "What is the best practice for research-based evidence to co-influence practice and policy?". In her response, she highlighted three important critical elements to influencing policy, namely: 1. the message, 2. the messenger and 3. the platform. These three (i.e. are detailed below), are central to development practitioners' approach.

1. **The Message**: Co-influencing is born from co-creation or co-production of the research-based evidence. During and after completion of research, there is a need for dissemination/communication of results—which is imperative. Important here is the quality of the message and noting:

   - the need to invest in processes, tools and approaches used for knowledge generation and/or acquisition which are inclusive.
   - that the message is not only generated through formal scientific research, but also capturing the voices, views, experiences and feedback from various groups of people that are affected by the policies.

2. **The Messenger (also referred to as policy champions)**: There is a need to ensure effective transfer of the message to go beyond preaching to the "converted". The "converted" here refers to fellow researchers by investing in developing capacities of strategic people to deliver the research evidence and bring the issues in the public domain and policy space. Some of the examples that can be used for effective policy advocacy include:

   - Training of farmers or communities to better articulating evidence (e.g., through use of theatre);
   - Training of researchers (especially young researches) on communicating scientific knowledge to be impactful.
   - Training of intermediaries (e.g., media, civil society) on interpreting and reporting on scientific results for wider reach.
   - Training of policy makers/policy bureaucrat on understanding, appraising (or valuing) research evidence.

Depending on the issues, these messengers could be drawn from different stakeholder groupings.

3. **The Platform**: Scientific papers and conferences by themselves cannot achieve required impact. In addition, the message and the messengers are not sufficient, if there is no "safe space" where dialogue for policy change can occur.

- There is a need to, therefore, pay special attention to creating engagement platforms where different actors have a 'safe space' for dialogue. Critical for the success and effectiveness of such engagement platforms, is to have the legitimate convening power and be trusted by all stakeholders. Other than face to face, other alternative platforms or products include electronic platforms (website and most recently, social media) and print (specifically, more formally policy brief and advisory notes).

Theatre for Policy Advocacy Guarantees Community Participation and Inclusion

Ownership of the developmental intervention by the beneficiaries is one of the major determining factors enhancing the sustenance of the project's impacts. However, most development efforts are decided in offices by highly experienced and qualified practitioners, who define the problems that communities experience; suggesting and designing solutions; setting up implementation dynamics; and ensuring monitoring and evaluation. Despite the purported participatory and bottom-up approaches, all this is done under the guise that the deemed beneficiaries of the planned development initiative—members of rural communities—lack the capacity to articulate their own challenges, as well as explore potential solutions. Whilst there may be an element of truth in these assertions, this somewhat condescending attitude results in developmental interventions that are disconnected from the targeted communities. Admittedly, it is a major challenge, but one that requires research and development practitioners to search for creative approaches to soliciting the input of community members at every stage of the project, from problem identification, establishment and prioritisation of possible solutions, implementation dynamics, monitoring and evaluation, as well as determining the role sort between the beneficiaries and project implementers. How then should a development practitioner harness the inherent potential and capacity in the communities where they target to intervene?

During the many years in AR&E4RD, I have learned at FANRPAN that Theatre for Policy Advocacy (TPA) is among one of the most effective approaches to addressing this challenge, to ensure that development initiatives take the beneficiaries on board. TPA is a tool that encourages the participation and involvement of ordinary people and communities in evidence-based policy advocacy with their leaders, service providers, policymakers; and other relevant stakeholders. Essentially, theatre is used to package and explain research findings and recommendations for behaviour change, and development practice and policy change by/to ordinary people and rural communities, as well as amplifying their voices to policy and decision-makers. The suitability of TPA to address the communications challenges in development stems from Africa's rich traditions. Story telling is core to the African culture, be it oral, literal or otherwise. For instance, in many African cultures, singing and dancing is a common practice, be it during work in the fields, at weddings, at funerals, and any other community functions. Because of its flexibility and connection with the key characteristics of

Africa's people, TPA can be used to communicate, advocate, educate, inform and serve as a trigger for soliciting responses on issues otherwise considered taboo.

TPA can be central to Social Behaviour Change and Communications (SBCC) and policy influencing messaging. For instance, in some TPA performance, whether in a conference hall or a village, I have seen people going home with new names based on the similarity of their behaviours with the character the Theatre presented. This has somehow assisted in the reinforcement of messages.

Conclusion

My humble beginning as a teacher and intern made me realise my interest to contribute towards solving societal problems as agriculture is purported to be part of the solution, especially among the rural small holder farmers, considering the differentiated landscape in the South African agriculture sector between post-democracy. To this date, the strong support foundation has guided my professional development and enabled me to progress to the level of a CEO of regional institute.

It should be noted that a career in development practices is not smooth sailing considering the personal sacrifices that one must go through. As part of this process, continuous training/ capacity building for individuals, systems and institutions to deliver lasting impacts is core. In addition, participatory approaches (such as the ARD, PEA and CI&I) have provided the development practice field with a lot of lessons, therefore, there should be a systematic assessment to gather these and be made available for consumptions to improve development practice. Finally, there is a need to bridge the gap between research-practice and policy making; and therefore, relevant skills and tools should be made available to development practitioners.

Tshilidzi Madzivhandila is CEO and Head of Mission for the Food, Agriculture and Natural Resources Policy Analysis Network's (FANRPAN). FANRPAN is an all-inclusive multi-stakeholder pan-African network that provides independent evidence to inform and influence policy processes at national and regional levels. With more than 25 years working experience, Dr. Madzivhandila is an experienced applied research, policy analysis, monitoring and evaluation, policy advocacy specialist in food systems, agriculture and natural resources. He has worked at the executive management level for six years. He holds a Ph.D. Degree in Economics—specializing in public policies and programmes evaluation—from the School of Business, Economics and Public Policy at the University of New England (Australia). He also holds a master's degree of Development Studies, Advanced Program in Marketing Management, and Higher Education Diploma and Bachelor of Agricultural Management.

Chapter 18
From Aid Recipient to Aid Manager—Living Life Full Circle

Mosarwa Segwabe

Introduction

Botswana gained independence from Britain in 1966. At the time it was one of the poorest and least developed countries in the world. The country was among those with the highest rates of poverty, with less than 20 km of tarred road, a poor health care system and a largely uneducated population and workforce. During that time, infrastructure of any kind did not exist. On numerous occasions, we were warned of impending post-independence disaster. Many had no to little glimmer of hope that the country would be able to stand on its own without the assistance of the former colonizers. Diamonds were discovered after independence. With good governance, the nation rose to become one of the richest and stable countries on the continent. Sir Seretse Khama, Botswana's founding President is credited for his prudent management of Botswana's mineral wealth, which paved the way for this enviable developmental success. These accolades followed all other Presidents who came after him in succession. Botswana became that seemingly elusive beacon of hope for Africa's development.

Over the years, Botswana used the revenues accrued from diamonds and other minerals to drive infrastructure development, extend free universal health care, significant social support system for the citizens and an education system that would be virtually free for all its citizens from primary to tertiary level. It is these deliberate efforts that have pushed up the literacy levels in the country, giving the nation a strong and healthy workforce for development. Today, every Botswana citizen is able to access a functioning health care system with a health facility within 8 km radius of where they live; something that was unheard of before and immediately following independence. Compared to other African countries, Botswana has a relatively stable political environment and has managed to hold free and fair elections every 5 years since independence. Having said all these, this landlocked country of just over 2

M. Segwabe (✉)
Gaborone, Botswana

T. Madzivhandila et al. (eds.), *Development Practice in Eastern and Southern Africa*,
https://doi.org/10.1007/978-3-030-91131-7_18

million people is far from reaching its full potential with regards to development; the economy is heavily reliant on minerals, especially diamonds, food production is very low with majority of food being imported, unemployment is growing and manufacturing is almost non-existent. These challenges present opportunities that need and must be explored to drive this country further up the ladder. The current stable political environment provides an excellent opportunity to entrepreneurs and others to explore the potential that exists in this country.

Memories from Developmental Aid Programs

Botswana has enjoyed support of many other nations around the world during the past decades. Here, I will share some of my memories. Growing up in my home village of Sefophe in the Central District in the early 1970s to the middle 1980s, I experienced the generosity of the American people. Botswana was still in its infancy in terms of development and depended largely on aid to operate. At the time Botswana probably received aid in several forms, however, the one that always comes to my mind are the many 50 kilogram (kg) brown bags of yellow sorghum and 5 litre (ltr) tins of vegetable oil that used to be delivered by big trucks either at my primary school or at the local clinic. As a young person growing up in the 70s, I came across these two very important items at two levels: (i) at the local health clinic as it was issued to all families who had under 5 year olds who attended the monthly child welfare clinic or to children who were malnourished and (ii) at school during break time as this same delicacy was served to all children at the primary school at 10 a.m. for the mid-morning break.

The chefs prepared the food outside using fire and big black pots. Every morning, each child brought either a bucket of water or a stick for firewood. This was necessary to ensure our serving of the delicious and good smelling yellow sorghum, during our mid-morning break. The yellow sorghum was such a delicacy. It was also very nutritious hence why it was also commonly prescribed for malnourished children at the clinic! It was cooked as a soft porridge and mixed with cooking oil. As young people, we enjoyed the meal that had a little bit more of the oil, where you actually see the oil moving in the plate and some sugar! To take care of this gap, most students would either request for sugar from their parents and more than half the time the request was denied, or they would resort to stealing the sugar. The extra oil and sugar turned the taste of the porridge into something to die for! What I can attest to is that every single person who ever had that meal loved it! It did not matter one's age or social status in life, we all seemed to love the meal.

Years later, as I reflect back on my primary school days, the memory that almost always comes back is the smell of that soft porridge mixed with cooking oil. It had a distinct smell that does not fade easily; the thought of this combination brings back wonderful memories. These two items came in special packaging; the yellow sorghum in a big 50 kg brown bag while the oil came in a four cornered 5 ltr silver tin. The two had a distinct brand that made them differ from all other bags and tins containing similar items. A big blue logo with a handshake, at the top of the logo it was written USAID while at the bottom there was an inscription that simply said, "From the American People". It was really years later that I got to understand why we ate that sorghum meal. And I came to understand why many of our schools, especially secondary schools had so many teachers that were simply called "Peace Corps Volunteers". These Peace Corps Volunteers were another form of aid "From the American People". My country was poor. My country benefited from aid. This aid and many other forms that came through other channels contributed immensely to the development of Botswana. One could say it was just sorghum and oil, but this sorghum and oil sustained hundreds and thousands of lives through their foundational years; child welfare clinic and primary school. Today most of these people are doctors, professors, politicians, Chief Executive Officers etc. Would they be in those positions of power if they did not get the nutritional value of the yellow sorghum? The generosity of the American people contributed to how Botswana was able to rise from being one of the poorest countries of the world to the Botswana that we see today, the Botswana that is an envy to many.

Many years later, I accepted a position as a Project Development Specialist with the United States Agency for International Development (USAID) in Botswana working in the area of HIV and AIDS. The first thing that was handed to me was a folder with orientation material inside. I looked at the picture on the cover of that folder and it brought back so many wonderful memories. Memories of eating the yellow sorghum and dripping oil combined with stolen sugar. The folder had the blue handshake with USAID scribbled at the top of the hand-shake and "From the American People" at the bottom. I felt like I had lived a full circle. From receiving aid that came from the American People through USAID and now working for USAID, with responsibility of managing this aid and ensuring it reaches the right people at the right time with the right services. The only difference is that now the Americans are bringing different types of aid; it is no longer yellow sorghum and cooking oil - how I wish!

Botswana, like most countries in sub-Saharan Africa, has been ravaged by HIV/AIDS. Through the US President's Emergency Plan for AIDS Relief (PEPFAR), the country is receiving funding towards HIV/AIDS prevention, care and treatment programs. I am at the centre of all this work, as part of the management team, with a full understanding of how it all works from when the funding is announced to when it reaches those that need the services the most and to reporting how the funds have been used. But, every time, I see that USAID logo, the smell of the yellow sorghum comes back. How fulfilling can life be?

National Service in Botswana—An Experience of a Lifetime

In 1980, Botswana introduced the Tirelo Setshaba National Service Programme. It targeted young people who successfully completed high school (Form 5) and obtained good grades (A and B). Participation in the national service was compulsory. Tirelo Setshaba (TS as it was popularly known) was non-military and involved individual placement of a high school leaver for 12 months in a different part of the country with a host family. The programme was designed to achieve nation-building, encouraged self-development of participants and enabled them to contribute to national development. Moreover, placing them especially in remote areas was meant to expose and sensitize them to social and cultural situations they were not familiar with. It is worth pointing out that serving in TS provided the youth with an opportunity to turn this time into an experience of employment and awareness of possible career choices for their future. During the service, a participant received a monthly allowance from the government for sustenance. The government put aside about 10% of the allowance that was then given to the participant upon completion of his or her service.

When this programme was introduced, I was still in primary school in my small home village doing standard four. At that time, I had no clue about this new government initiative that would later touch my life in considerably positive ways. At that age, my main concern was family (immediate and extended), especially my parents, siblings and friends. Friends were a big deal, especially when we came together after school and on weekends to playhouse using anything and everything old that we could lay our hands on. We used old bricks and stones to make houses. Old tins were harvested and used to make pots. We cooked leftover food when we were lucky to stumble on some. However, most of the time, we 'cooked' sand to make up for the food. Those were the good times. Life was simply beautiful.

I was always a good student in terms of manners, behaviour, conduct and school performance. I was also one of the go-to-learners when a teacher needed an assistant in class. There were times when I was sent to their homes to bring breakfast or pick a book they might have forgotten. I completed primary school and was admitted at a boarding school. Fast forward to 1988. I wrote the high school (Form 5) final examinations and I obtained one of those grades that qualified one to be a Tirelo Setshaba participant. As noted earlier, it was compulsory for anyone who obtained Grade A or B. Anxious and scared as I was, I had no way of getting out of it. Otherwise, I was not going to secure a scholarship to enable me to enrol for a university degree. This was prestigious then. Everyone looked forward to entering university after passing Cambridge examinations. Most of my friends were happy and excited about the possibility of gaining independence and being on their own. In contrast, I was extremely uncomfortable and anxious about the looming unknown. All my life, I had always lived with my parents. At no time had I slept in a room alone. Nor had I ever found myself planning for my meals and other household responsibilities. My parents were always near. Even when I was at boarding school, my parents lived 27 km away and regularly came to visit. They ensured that I had the basic supplies I needed all the time. Here I was. Now it was my turn to do it all by myself. What was it going to be like? I wondered.

My journey to self-discovery and reliance started in April 1989 with a week-long orientation programme in the central district of Botswana, a 2–3 h drive from my home village. That counted for something and brought some comfort because it meant that my final placement was likely to be not too far away as I had thought earlier. My elder brother for example, was placed in a very far away place that required spending an entire day on very bad roads to reach his destination. This was what was mainly known about the programme. Most participants were placed far from their families to create opportunities to learn about and embrace other cultures in the country. It was anticipated that by so doing, we would nurture a sense of one Botswana. After the orientation week we were divided into sub-groups and boarded trucks. We travelled to different sub-regions where we underwent further orientation. Finally, after another week of additional orientation, we were loaded into trucks again and dropped at what were to become our new villages, homes, families, friends and colleagues for the next 12 months! The real journey into self-discovery and learning about life was now beginning in earnest.

The village I was placed in was about a 4-h drive from my own home village. May be one can say it was not so far really. That may be the case. However, it was very different from the one I was raised in and all other villages that I had seen and visited by that time. First, the people there spoke Kalanga, which was a totally different language altogether, which I could not speak or understand. Homes were scattered all over with huge swaths of land in between. Mopane trees filled the open spaces, which made it difficult to see the nearest homestead. That was so different from the situation in the village I came from. The school, clinic and kgotla (village parliament that the Chief chaired) were all in one place. To get to these places, I had to walk for about 45 min through Mopane tree bushes and cross a river. One main tarred road passed about 10 km outside my village. There was no public transport to take one to the main road. Getting there meant that I had to walk or wait for several hours to days to get a ride. Once at the main road, it was easier to get transport to one's destination. This information was not so useful to me because as a TS participant, I was not allowed to leave my village. The only exception was during certain times, for example to attend a funeral of a family member or when going home for the Christmas holidays. At the end of my 12 months, I had learnt that humans as social beings had innate abilities to adjust and rise above all challenges that might be thrown at them.

I was lucky to have been placed in a wonderful host family. They gave me a whole round hut to turn into my house for the entire 12 months. I remember on that very first day I arrived, it was very clean, which pleased me so much and confirmed that I was welcome. My heart skipped a beat though when I moved the curtain to open the window and realized that one of the windowpanes was missing. I freaked out. All I could see from that window was a stranger (most likely a man) putting his head through the window and harass me. That fear forced me to reach out to my host mother. She totally understood my situation and offered my host sister to share the room with me. I gladly accepted the offer.

Yet another surprise was waiting. I realized that there was no toilet on the homestead. My brain started wandering all over. Will I be using the bush? I was told to use the toilet at the nearest store, which was about a five minutes walk from my host parents' home. I immediately worried about what I would do at night. Fortunately, all in all, my host family was very accommodative. Even though our main challenge was language, members of my new family tried as best as they could to make it easier for us to understand each other. I also tried the best I could to complement their efforts. Over time, I learnt a bit of the Kalanga language. I must admit that I did not find it an easy language to learn though. By the time I completed my tour of service, I was able to follow conversations in the language as long as people involved spoke slowly. At that time, I could say a few words. What a huge learning experience it was for me! It dawned on me better that we were all Batswana, though different in our ways. Our languages, lifestyles and cultural traditions were different. Despite all this, I learnt during that period of living with this family that our family values as a nation were the same irrespective of which corner one came from. I discovered that the pillar of the Batswana heartbeat was being caring, loving, accommodating and nurturing.

Tirelo Setshaba did not only create opportunities for young people straight from high school to live in a different part of the country with people with a different culture. It also gave them a chance to acquire skills and experience across various industries and business sectors. These opportunities were beneficial to the youth and contributed to economic development of their communities. In my small village, I worked with various government sectors, among which were education (primary school), clinic, customary court (police and chief) and agriculture. I worked on a rotational basis. One week was spent at the school as an assistant to the Standard 6 teacher. The following week saw me being a Nursing Assistant at the clinic. There was no nurse at that clinic. Next was an opportunity to be with the police. Lastly, I would work with the Agricultural Officer before the rotation started all over again. I might not have chosen any of these professions eventually, but I remain grateful for the unique and privileged exposure. Through that placement, I learnt how to offer proper customer service and what it means to be accountable and responsible. I cannot forget the experience I had when I saw firsthand, teachers with very limited resources doing amazing things to ensure their learners covered the content they were expected to master. I saw the Nurse Assistant go above and beyond her normal duties to ensure the people of her village received the best healthcare ever. How can I forget the Police Officer who worked with little to nothing to maintain order, safety and security in the village?

At first I was anxious and afraid of what to expect when I joined the national service. Twelve months looked like a lifetime. Yet, when I look back now, all I can express is extreme gratitude for all that I had learnt, the wonderful people I met, rich memories I archived and life changing experiences I had accumulated. Let me point out that I am still grateful that I took good care of myself. I was able to use the monthly allowance I received responsibly. At no point did I go to bed on an empty stomach. I returned home after 12 months ready and prepared to start a new chapter. The only difference this time was that I was moving to the capital city to embark on my 4-year journey at the University of Botswana.

Professional Growth in the Era of HIV/AIDS

The first case of HIV/AIDS in Botswana was diagnosed in 1984. At the time I was doing form 1 (Grade 8) and just starting to learn about HIV. During those times, most of our news was coming through the radio and newspapers as there was no internet or social media and access to the television was quite limited. This meant news took longer to reach people. But what I remember the most were stories about places like Uganda being the epicentre of HIV/AIDS in Africa. At the time, Uganda seemed very far and it never occurred to me that this disease that seemed so deadly would one day wreck my country in ways never imagined, before reversing all the gains the country had enjoyed since independence in 1966. It also never occurred to me then that fighting this disease would come to define my professional life. Indeed, life is full of the unknown.

I started becoming actively involved in the fight against this epidemic during my under-graduate years at the university when my country was starting to experience the full brunt of HIV and AIDS. I volunteered for the Social Work Students Against AIDS peer education group; to educate and create awareness among students—that was the beginning of my involvement with this epidemic that later ravaged not only my country, but the entire sub-Saharan Africa region like nothing ever seen before. HIV/AIDS crippled countries' economies, challenged health care systems, separated families and left multitudes of orphans and vulnerable children behind. While at the university, my experience with HIV/AIDS was still limited to what I heard, read and learnt in class. I still had not really seen the ugly face of this disease yet. And at that point I still had no idea of what I was about to experience.

It was after I completed my undergraduate studies, fresh-faced, and ready to take on the world that I stepped into my first job at Nyangabgwe Hospital in Francistown, Botswana as a social worker. What I encountered was a nightmare. No time was wasted on orientation and easing into the job. Immediately I was faced with death day in and day out. There was no treatment. Lives were lost every minute and families were devastated. The health care system was stretched to the limit. Those are difficult days to remember and ones I never really want to spend too much time thinking about as the thoughts bring back a lot of pain. Our patients were dying every moment and we were helpless. The children in the pediatric medical ward broke my heart the most. I had several nightmares; pictures of these children on life support struggling to breathe. I came across young mothers who fell pregnant, tested HIV positive and had no idea where the father of the child was. Some young girls diagnosed positive continued with their lives normally no matter how much they were engaged, especially those that did not show signs of being ill. Experiences with all these people broke my heart.

One Saturday evening, I was home with my cousin who was visiting me, and I had planned a sleepover with her. Suddenly, my beeper went off. I checked it and it showed the number from the pediatric medical ward. As soon as I picked up the phone to call the ward, a car hooted and when I looked outside, it was the ambulance—to pick me up. I was needed at the ward because a grandmother wanted her 2-year old grandchild to be discharged against medical advice. She wanted to take the child to a traditional doctor because she believed the child's front fontanelle was not appropriately treated the last time, she took the child to see the traditional doctor. I spent over three hours with the grandmother educating her while also trying to understand her point of view. She was adamant that the best thing to do was to discharge the child, get the child treated traditionally first and bring them back to the hospital immediately after the traditional processes were completed. While our intention was not to take away from her what she believed in, we could not discharge the child. It was an extremely delicate situation that needed to be handled with care. It was heart-breaking to see the grandmother break down and cry. What do you say to her? How do you strike a balance between traditional and modern medicine when dealing with life and death issues? The baby was on oxygen and we knew that as soon as we took him from the care he was receiving, he would not make it. Even if he was going to make it to the traditional doctor's place, how do we know what he was going to be given, how much of it and for how long and how would we have known how the traditional medicine would interact with the modern medicine the baby was already on? I left the hospital at midnight to go home and sleep with the baby still in the ward. Unfortunately for us all, the baby passed away a few hours after I had left. Later the next morning I went back to the ward to check on my patient and the bed was empty. The nurses who were on day duty received a report from those that were on night duty that the grandmother became hysterical, running all over looking for the little girl who refused to discharge her grandchild from the hospital, the little girl who killed her grandchild. The little girl was me. My heart broke into pieces, not because I was referred to as a little girl but because of losing yet another pediatric patient to HIV and AIDS and because of the pain and suffering that yet another family had had to endure.

Unfortunately, this was not the last heart-breaking experience I lived through. Overtime they became too many to count. There were many Monday mornings whereby mid-morning my in-tray would start to fill with new referrals from the different wards. Over 90% of the referrals would be HIV/AIDS related. I was so used to the routine which included: Do pre-test counselling. Do post-test-test counselling (HIV+ result). Get in touch with the family and prepare them for home-based care; the patient is HIV+. Abandoned child found in the bush admitted in maternity ward; blood sent to lab for HIV testing. 3 people died over the weekend; failed to get in touch with relatives listed in the medical card and we suspect they are illegal immigrants-please trace relatives and arrange for repatriation of bodies. In the meantime the man who was in-charge of the mortuary would be calling me, telling me about the body of an unknown individual suspected to be a non-Motswana who had been in the mortuary for more than 6 months and how the body had started to decompose and something needed to be done. After going through all the referrals and prioritizing which one to see first, I would go to the wards to make follow ups. Most of the time, I would get to the bed and find it empty. When I asked the nurse where the patient was, the nurse would say, did you meet a porter on the way pushing the navy-blue trolley to the mortuary-that was your patient. This turned out to be normal for most days of the week. It was draining. With time though, I started being numb. Delivering HIV positive results became normal. Death became normal. I went home and I slept. I worried less unless the diagnosis was for someone I knew closely. It was too much. While it seemed "normal", it was actually abnormal to feel that way.

I left the hospital environment in 2000 but still worked in the field of HIV/AIDS focusing on voluntary HIV counseling and testing, health education and health promotion, research and sometimes teaching at the University of Botswana. I continued to deliver HIV+ results to patients and also see death, but nothing close to what I saw in the hospital environment. In 2003, the government of Botswana started a free national anti-retroviral drug program. I remember my excitement to the news. I was still in the middle of this fight. I immediately got engaged in the new program, leading a team that educated students and staff at the university of Botswana about the new treatment program. Dawn had arrived. After losing thousands of people, seeing increased numbers of orphans and vulnerable children, increased child headed households, families breaking down and apart and unmeasurable sadness covering our nation, a new day and brighter future started to unfold in front of us.

Over time, more and more people enrolled on the program and death rates went down. People that were confined to their beds started getting back on their feet and even going back to work. A new normal started to unfold in front of us; the normal with fewer funerals every weekend. The normal with more and more healthy-looking people out in the streets going on with their lives. The normal where mothers could enjoy being mothers and not spend their entire time caring for one sick child after another. The normal where the future became brighter and brighter; that brightness still shines today. Remember I referred to the young mothers I met while I was working at the hospital and how they did not care much back then? Well, I did meet one of them later in life--several years later. And God had spared her to also see and live in this new dawn. I hardly recognized her, but she recognized me even when I was in the car. She immediately came to me and confirmed my name. Upon my response, she threw herself all over me, very happy to see me; she just could not contain her excitement as she told those that she was with who I was. And she said, do you know how old your child is now? Referring to the child she gave birth to while I was still working at the hospital. She gave me a quick update. The child had grown into a fine young man and at the time was at the university pursuing his undergraduate degree. She thanked me tremendously for saving not only her life, but her son's too. They were both on treatment and doing well. They were living the new normal; basking in the dawn brought about by the HIV/AIDS medication and she was proud to share her story.

Conclusion

These lived experiences have shaped my life. Reflecting on these experiences has once more reminded me about the rich life I have lived. The stories point to my personal development of a little girl growing up in a small rural village of less than 5,000 people. A little girl who grew up and maneuvered through life, ending up living a life of purpose made very rich by the unique experiences she acquired along the way. I trust readers will find these stories educational, motivating, encouraging and thoughtful. I also trust the reader sees that no matter how difficult life can be, that there is always a light at the end—one has to believe and hope that a bright future always lies ahead in the horizon. That we are all here for a purpose and that one of them is to serve others.

Mosarwa Segwabe has over 25 years of experience working in the fields of HIV and AIDS and health and wellness as a social worker and public health professional. She has worked in the health, social, university/academia, private sector and donor environments in Botswana. She has spent many years managing health education and health promotion programs, prevention, care and support programs for orphans and vulnerable children and youth in Botswana. Her broad experience also touches on the areas of monitoring and evaluation including conducting research in different areas of health focusing on HIV and AIDS and its impact on orphans and vulnerable children especially, with some of the work ending in publications. She was born and raised in Botswana. She obtained a Masters in Social Science Administration from Case Western Reserve University in the USA and Master of Public Health from The Medical University of Southern Africa (MEDUNSA) in South Africa.

Chapter 19
A Cocktail of Sorrows and Joys from the Trenches of my Development Practice

Joseph Francis

Introduction

Approximately 265 km to the north-west of Harare, the capital city of Zimbabwe, lies my rural home. It is in a village that falls in an area under Chief Mzilawempi in Hurungwe District of Mashonaland West Province. My close relations and older people often tell me that I was born in a hut on a former large-scale White-owned commercial farm near the then Mhangura Copper mine. However, the foundations of who I am now were defined and incubated in Chief Mzilawempi's area.

I was told that my parents' marriage disintegrated when I was barely three years old. Soon after this unfortunate development, my widowed maternal grandmother took custody of me and my two sisters. I became the oldest male member of the family that my grandmother raised in Nyarumwe village. My life in the village was characterized by both joyful and sorrowful experiences and became the pedestal on which I stood. Over the years I navigated various terrains to become the Professor of Rural Development that I am now. As I reflect on my life, I marvel and appreciate the numerous players who nurtured me for more than half a Century of my life. This and my 28 years of marriage also contributed a cocktail of experiences and lessons that will remain permanently inscribed in my mind and professional practice.

Now that the "trench" is partially known, nothing excites and makes me prouder than sharing my reflections on my unique journey in development practice. The experiential stories that I have shared below are meant to help construct a resource that contributes to showcasing the realities of development practice in eastern and southern Africa. The stories I am sharing are structured as follows: (1) Grandmother's perspectives on love: can this be a solid foundation for sustainable development? (2) Seniority in a development-oriented organization does not translate to knowing it all; (3) What a conspiracy of silence by smallholder dairy farmers! (4) Her kick below

J. Francis (✉)
University of Venda, Institute for Rural Development, Thohoyandou, South Africa
e-mail: joseph.francis@univen.ac.za

T. Madzivhandila et al. (eds.), *Development Practice in Eastern and Southern Africa*,
https://doi.org/10.1007/978-3-030-91131-7_19

the belt that almost destroyed my appetite for development work; (5) Engaging men at drinking havens to secure support for development work; and (6) Two minutes that unlocked respect for our engaged work.

Grandmother's Perspectives on Love

Quite often we become exasperated when confronted with tensions and conflicts in grassroots communities. The frustrations emanate from the difficulties we face as we strive to successfully implement development initiatives. As development practitioners, we try our best to understand the situations and develop solutions but we often fail to achieve our targets. Perhaps our upbringing within the context of a weak family structure engenders this failing. Here was an old lady who had never been in a classroom all her life. Yet, she had so much wisdom that I believe can shape how we build bridges within and across families, leading to sustainable communities. Her narrative with respect to "love" is so profound that it deserves special attention among her numerous teachings. For her, love must be the central pillar of stable families and communities.

I recall one day when we were enjoying a special traditional meal that my grandmother had prepared for dinner. As had become the norm, especially when she prepared the special meals, we always sat in our grass-thatched rondavel that served as our family kitchen. It was an unwritten norm for her to prepare a special traditional meal for the family while she enjoyed her self-brewed beer. By the way, she was highly regarded far and wide in the villages in Hurungwe for brewing the best traditional beer. As she drank more of her beer, she shared many stories about things that she had experienced in life. We enjoyed her well-articulated lived experiences that she readily narrated. Then came the most intriguing moment when she posed the question, "what does love mean to you?" Just imagine how we tripped over each other to express our views.

As we took turns to answer the question she posed, laughter was an ever-present companion. Upon noticing that we seemed to have exhausted our various perspectives, she then took the centre stage. First, she conceded that it was difficult to provide a simple definition or explanation for "love". For her, love started with the individual "being" not doing anything that would invite harm, ridicule or shame to himself or herself. This entailed cleaning one's mind and mindset to eliminate any lingering negative thoughts or plans. In addition to this, as an expression of self-love, one should bathe regularly and wear clean, ironed clothes. Moreover, love meant that the individual always maintained positive relations with members of his/her family. All actions towards others should reflect warmth, respect, trust and that ever readiness to assist where possible. She stressed that social giving should be one of the most cardinal values that characterised love amongst members of our family. Food, no matter how little was available, was supposed to be shared even with neighbours or strangers present when our family enjoyed its meals. Her regular injunction that remained indelibly printed in our memories as we grew up was that we should "never hide food when you see visitors approaching our homestead. Instead, offer them some of what the family might be about to eat". My grandmother believed that beyond our immediate family, we should always treat those from outside in the same ways as those closest to us. In this vein, she emphasised that anyone who seemed to be the same age as her was also our grandmother. The same was the case with our brothers, mothers, fathers, sisters and other categories of family members. Although some of us had reservations about this narrative, we found ourselves complying. As I reflect on this now, I can only appreciate her immense wisdom and priceless lessons with respect to how to build social capital, which ultimately cements a sense of belonging and bedrock of a sustainably flourishing community.

Given the decay we continue to witness in family institutions and associated implications in our struggle to improve people's livelihoods, can it then be that my grandmother's perspective on love is a recipe or ingredient for reconstructing a stable society and unity in a country? And can this be the answer to how development beneficiaries behave when we visit their homes to collect data? If the responses to these questions are in the affirmative, what might it take to realise this reconfiguration? Can this not be the answer on how to construct strong families and communities, leading to sustainable development? Furthermore, do these narratives bias the data collection responses we receive (for instance, from those of similar inclination to my grandmother's line of thinking who are more often the matriarchs of today's rural society). This is distinct and significant in rural settings.

Seniority in that Development-Oriented Organization does not Translate to Knowing it All

I still remember a day in September 2002 when we flew from OR Tambo International Airport in South Africa to one of the neighbouring countries. We were on a mission to launch an integrated rural development programme that we were implementing in six countries of southern Africa. I was only three months into a Postdoctoral Fellowship in this development programme, in addition to being the only male member of the delegation.

A day before we left on this trip, our Regional Director and Programme Manager organized a meeting to reinforce and finalize our plan for the official launch of our programme. We discussed many issues that included local protocol issues, dress code and centrality of being punctual, among others. Having just joined the development programme, I found it was prudent not to say much in the meeting. However, as the meeting progressed, I just could not remain caged in my shell. Thus, I found myself contributing a few ideas that I believed had not been treated with the level of importance and seriousness they deserved. First, I indicated that given the strong cultural traditions and beliefs of the implementation site we were going to, it was important to seek guidance on many issues from the local organisers of the event. Second, I suggested that if the ladies had traditional attire, they should prioritize wearing that over any other types of regalia. I cautioned my colleagues against putting on pants (trousers) and highlighted the need for preparing for other unknowns. Use of headgear and carrying a piece of cloth large enough to be used should it be necessary to sit on the floor during proceedings was not far-fetched. Third and lastly, I lamented the fact that we had not travelled at least a day earlier given that we opted to fly out on the day of such a crucial event. As the discussions unfolded, it was clear that my colleagues were not fully in agreement with my suggestions. Having read in between the lines, I retreated into my shell again and held my peace. At least I had made my views known. Eventually, time would tell.

On the travel date, we drove together to the airport, checked in well ahead of our departure time and proceeded to wait in the departure lounge. As we waited, each one of us minding his/her business, things started to unravel. It was announced via the public address system that our flight would be delayed by almost an hour. At that point, a lot of things started playing out in my mind. Had it not been revealed during our meeting the previous day that it would take about 1–2 h to get to the venue of the official launch of our development programme? I estimated that we were likely to be late by at least half an hour. Had I not raised the importance of travelling at least a day earlier? I kept all these thoughts to myself. Even though no one expressed it openly, I could read that anxiety (and perhaps sanity) had invaded my colleagues' minds. Eventually, we flew out and arrived at the launch function venue about an hour late. One of my fears had been confirmed yet again. Considering our late arrival, was there going to be any briefing session with the local leadership as originally scheduled?

About 200–300 people were waiting at the open space where the event was to be held. The people were already seated, seemingly without anything official in progress. There were no ushers in sight. The men sat under a pitched blue tent on one side. Women sat on the floor facing the pitched tent. I walked in front of my colleagues, aiming to somehow manage things but without saying so. Based on the experience I had acquired in the past, I was aware that one should not just find the nearest seat and occupy it. It was wise for us to wait and be shown where to sit. I stood and waited to be directed to my seat and expected my colleagues to follow suit. However, what followed thereafter was a scene from a real drama script, which not only embarrassed our team but was a huge lesson.

Our Director briskly and confidently walked past me. She occupied the first vacant chair among the men in the tent. Evidently, she was carrying her seniority in our organisation and expecting to be treated as such at this meeting. Call it a clash of cultural traditions and 'modernity'. Meanwhile, I kept monitoring the body language of the audience. It did not take long to confirm that a serious miscalculation had been made. Would this end well?

Many women gestured feverishly to our Director as they tried to draw her attention, inviting her to join them. Seeing that their gestures were not yelding their desired result, one elderly woman stood up and proceeded to grab our Director by the hand. She virtually dragged her to join the other women. Another woman offered her own cloth for our Director to wrap herself with. Incidentally, she was the only female putting on a pair of trousers. That was not the end of the drama. What a spectacle!

As the drama unfolded, a procession of about 10 men emerged from behind the tent. Everyone stood up. Women ululated. Men whistled and said a lot of things in praise and reverence. I could not understand what they were saying. However, it was evident that those were words that praised their leader. I tried to join in without success of course. Don't they say, "when in Rome, do as the Romans do?" That was the grand arrival of the Chief of the area.

I was curious to see and read through the confusion that enveloped our Director as all that was unfolding. After all the drama that had just unfolded in our eyes, we were directed to our respective seats. Mine was among the men. My colleagues joined other women. Thereafter, the ceremony proceeded without any further unsavoury incidents. Nevertheless, I listened attentively to the Programme Director and one speaker after the other. At least our Director apologised for all the mishap at the beginning of the proceedings. That was a clever move on her part. What an educational platform and learning expedition that was! What a learning experience for our leader of delegation and others of similar persuasion! Yes, the experience was unpleasant but this was a significant reality check.

The lesson from this story is that occupying a senior position in an organization does not translate to being the smartest brain or fountain of knowledge on development matters. Titles or positions of authority get recognised on platforms that are appropriate. Unfortunately, this was not one such platform.

What a Conspiracy of Silence by Smallholder Dairy Farmers!

Informed decision making in development programming is best if the information before us is current, relevant and authentic. From 1995–1997, I led a research project that also included training of smallholder farmers in a dairy development scheme in Zimbabwe. It focused on how to keep and utilise mixed crop-livestock farming enterprise records. Thirty-two farmers in both small-scale commercial and communal farming areas participated in the project. After active engagements at individual and group levels, the 32 farmers had volunteered freely to participate in the project. Each farmer maintained a set of records captured daily, including income and expenditure profiles. At the beginning of each month, the farming households received a printed sheet, which they used to record the money they received and spent in their dairy enterprises. A full-time field technician visited each farming household at least once in a week. On each visit, he checked the correctness and completeness of each farm's records. Whenever he identified or was informed about issues that needed attention, the technician was expected to address them.

What surprised us was that almost all the farmers consistently asked for the income and expenditure record sheet a week after we had delivered to them one. I still remember vividly that this happened for three successive months. Although we tried to understand what was going on, it was all to no avail. Eventually, we got the answer during our first quarterly project review meeting with representatives of the participating farming households.

One elderly farmer stood up during the project review meeting. He revealed and explained that most of the heads of households participating in the project had agreed to maintain two records of income and expenditure. In one of the sheets, grossly under declared income and expenditure details were recorded. The latter record was meant for my research team. At the same time, authentic details were maintained in another record sheet that they kept as confidential information. This unfolded until the farmers were confident and convinced that there were no ulterior motives behind what we were doing. All along, this was the reason why they asked for a second record sheet that enabled them to keep the two sets of information. The old man went on to point out that the farmers were now convinced that our intentions were noble and developmental. Soon after this explanation, all the farmer representatives pulled out and handed over the authentic parallel records, which they had maintained for the past three months. We were astounded, perplexed and amused that we had been dribbled past in such a deft manner for three months. What a revelation, experience and challenge for development practitioners and community-based researchers!

Up to now, I remain amazed at how the farmers had sustained their conspiracy of silence. What made it possible to maintain this veil of secrecy for three months? Given this sobering experience, are once-off questionnaires appropriate for collecting reliable data to use in rural development programming? Could this be the reason why most rural development programmes often fail? Thus, how much time should we invest in nurturing the development of trustful relationships between external agents of development and grassroots communities?

Her Kick Below the Belt that Almost Felled my Appetite for Development Work

In 2018, the South African government adopted a the "National Framework for Local Economic Development: Creating Innovation-driven Local Economic Development (LED), 2018–2028". Later that year, I successfully won funding from a Government Department to implement a national project that pilot-tested how to integrate innovation into LED in one District of the country. I resolved to co-lead this sole national pilot project in the country with the LED Office of the District Municipality. This was the first time I had secured considerable funding for a local government project. In addition, this was the first time the government Department that funded our project had ever invested in that District. I expected my colleagues in the Local and District Municipalities to appreciate this significant development that was also meant to demonstrate the relevance of my university in societal development. If only that was the case! Let me share one of my many experiences that almost convinced me that I was working with wrong people.

Soon after the project was approved and before receiving the first tranche of funding, I embarked on an awareness creation, orientation and stakeholder mobilisation exercise. In this journey, I made sure that I made and implemented key decisions with the LED Manager of the District Municipality and also with the knowledge of the General Manager responsible for development planning. To secure support for the project, we presented the project to the District Municipal Mayoral Executive Committee together with the General Manager and other officers in her Department. All this was in sync with the belief that initiatives of this nature must be co-designed, co-implemented, and co-monitored and evaluated with integrated co-learning. This process was to ensure co-ownership and mutual accountability amongst all the stakeholders, especially from the District Municipality.

It was not long before questions regarding how much of the project money would directly be used or disbursed into the coffers of the Municipalities started emerging, though said in suppressed tones at first. No matter how much we explained to the municipal authorities the nature of the project and what the funding covered, this matter could just not go away. Matters boiled over when we held a special meeting of the Project Steering Committee (PSC). In our first ordinary meeting, I had been tasked to prepare a document that articulated the theory of change for the project. The special meeting was arranged to deliberate on this document and possibly adopt it. The Director who headed the Directorate in the government department that funded the project chaired this special PSC meeting. Other persons in attendance were representatives of key departments of government that were relevant to the implementation of LED.

The constructive comments that followed my presentation confirmed that the participants received it well. Unfortunately, this positive atmosphere failed to dissuade the General Manager from the District Municipality to whom the LED Manager reported from pouring cold water over all of it. She had arrived at the meeting when I was midway into my presentation. When she walked in and joined us, I thought her presence would add value and weight to what we were doing, especially in the eyes of the funder of our project. How wrong I was!

When given the opportunity to contribute, she made several trivial comments and released a flood of questions that almost derailed me. Out of ignorance, or for whatever reason, most of her questions had been covered already in my presentation. Having joined the meeting late, she had not heard what had been said. In an apparent fit of rage, her complaints included why we had excluded the logo of her Municipality among those of institutions leading the project. After a relentless ranting about so many other things, some of them not so related to the project, she let go of the most disgusting "kick below my belt" attack. She wanted to know whether the project was meant to collect data for my postgraduate studies. Seemingly, according to her, as a Professor of Rural Development I was embarking on a research towards a postgraduate degree! To say that this was annoying, and anger-inducing is an understatement. In the face of all this frontal attack and provocation, I remained composed as I prepared my response. My response was not a counterattack because I realised that this was the best time to adhere to the advice enshrined in the saying, "Never roll in the mud with a pig, for people might not notice any difference". My suppressed anger helped me maintain a semblance of sanity as I answered her trivial questions while also addressing her openly declared concerns.

It was interesting to note that the General Manager had incensed many other participants in the meeting. They called her to order and released the cat out of her political bag. The situation almost boiled over, which made me request the chairperson of the meeting to take a break, which took about 15 minutes. A few of us used that time to engage and make her apologise for her conduct, which was so damaging. What was even more significant, and damaging was that the funder went to the extent of contemplating taking the project to another district. With this threat hanging over the head of the project, the post-meeting period became an extremely challenging one. We tried the best we could to assure the funder that despite the uncouth behaviour of the General Manager, the project was still in good hands. I am happy to say that we are still running the project as originally planned. In addition, smiles are punctuating implementation of the project.

Considering the behaviour revealed above, is it surprising that the Auditor General classifies more than 80 % of Municipalities in South Africa as dysfunctional? What chances exist for us to achieve the sustainable development goals given the calibre of the General Manager in this story who behaved like a mischievous and misguided politician? What else could we have done to reduce the chances of the unfortunate scene witnessed in our project meeting?

Engaging Men at Drinking Havens to Secure Support for Development Work

From 2010 to 2012, I led a project that focused on unearthing what the mantra, "The people shall govern" meant for residents of a Local Municipality in Limpopo Province. We held a series of meetings and workshops with the aim of creating awareness and securing support of various stakeholders for the project. Six months after introducing it to the Wards where we were implementing it, it became clearer that men were worryingly apathetic. Thus, we resolved to engage the men wherever they spent most of their time.

A casual investigation revealed that most men spent considerable time at specific drinking spots. Among these were local taverns (mainly liquor trading stores). Thus, I decided to spend some time at some taverns even though I never drank beer. Let me share one of my exploratory engagements.

Although I was now well known in the area, no one expected to see me at a tavern that afternoon. It was not surprising that eyebrows were raised when I spent about an hour at the local tavern. I bought a crate of beer, with the assistance of a member of the local community I worked with. I took the beer to a spot with a group of men drinking. As old as it might seem, I regarded this approach as a way to unlock insightful, frank and sustainable conversations. Whilst they enjoyed the beer, I was downing the orange juice I had bought for myself. They were warming up to me even more as we discussed various topics, ranging from local issues to international affairs. Among the group of men were two agricultural extension officers.

It was time to say my goodbyes for the day. Reluctantly, they let me leave. One of them requested that we take him home given that it was getting darker. His home was along the route that took us back to where we were booked for the week. He indicated that he would have walked back home alone. Without any hesitation, I welcomed him to join us. What surprised me most then were the glances the men he was drinking beer with exchanged among themselves. I suspected that there was a story behind that reaction. We were to find our answers from our passenger as we drove back to our hotel.

Our passenger revealed that virtually all external agents of development who worked in the area, including some government officers, never acceded to requests for transport from the locals. Furthermore, he indicated that our readiness to agree to his request, our punctuality when meetings were held and general behaviour that was "not the same in comparison with other university researchers who worked here in the past" were attributes that made the project interesting, appealing and likely to succeed if sustained. We took note of this advice to heart. Why not?

In the days, weeks and months that followed, we got one piece of unsolicited advice after another regarding how the various stakeholders in the area could be brought on board. New acquaintances and friendships were cemented. Some of the relationships established then are still intact even today, which is more than two decades after the project was completed.

The experience I have shared above makes me wonder whether we ever value spending time just talking to people in their favourite places of socialization to gain insights into the psychology of community life? Such excursions in development practice are evidently the pillars that yield trust, respect and social giving for successful implementation of development work. But how prepared are funders to accept inclusion of such aspects into project plans? If not, what must we do to influence them to consider this?

Two Minutes that Unlocked Respect for Our Engaged Work

Every year, the Talloires Network[1] invites member universities throughout the world to submit community-engaged projects for consideration for award of the MacJannet Prize of Global Citizenship. The prize is awarded to recognise and encourage exceptional engaged work in which students are actively involved. Late 2010, a Professor at a university in the USA advised and encouraged me to submit our "Amplifying Community Voices" project for this award. At first, I was reluctant to prepare and submit the application. However, after careful consideration, I successfully applied two days before the closing date. At that time, senior managers and academic staff

[1] www.Talloiresnetwork.tufts.edu.

in general in my university regarded community-engaged work as an unnecessary nuisance because they believed it was not 'science' driven.

The situation I have articulated above prevailed in an environment where the engaged work I led often attracted 150–200 students every year. They were mainly undergraduate students from various Schools of the university who championed the "Amplifying Community Voices" project activities. We carried out the work in more than 110 villages in one District Municipality of Limpopo Province. Our work entailed students facilitating community-based workshops in which local citizens deliberated on issues that affected their lives and livelihoods. Moreover, the students organised various activities to celebrate or commemorate national and international holidays. Their work, for which I was the mentor, placed most emphasis on developing responsible citizenship within the student community as they created awareness or educated grassroots society on various contemporary development issues. Facilitating community development planning through harnessing participatory action research tools as they engaged children, youth, adults and local leaders to uncover their perspectives was the most popular. This work was also so popular that over time we received numerous requests to extend it to other communities beyond the District Municipality in which our university is located. This is the work that underpinned our application for a possible award of the MacJannet Prize of Global Citizenship in 2011.

Towards the end of March 2011, it was officially announced that out of the 75 applications that were received from 59 Universities in 26 countries the "Amplifying Community Voices" programme had been ranked third. This was a significant development for us because for the first time our university had won such an award. We were informed that we would receive the award in June 2011 at an international conference to be held in Madrid. That conference would bring together Vice Chancellors or Senior Executives from Talloires Network member universities throughout the world. The organisers of the award ceremony advised us to prepare an acceptance speech to be delivered in two minutes, which we did. Our own Vice Chancellor confirmed that he would attend the conference and award ceremony. One of the students participating in the project was supposed to accompany me to receive the award. I asked the students I worked with to meet and recommend one of them who they believed could be their most suitable representative at the award ceremony.

The prize giving day, which was held on 14th June 2011, arrived. We were in the hall where the more than 200 delegates and invited guests were gathered for the ceremony. Vice Chancellors from 13 of the 23 South African universities at that time, including our own, were in attendance. All the formalities were completed. It was time for the prize winners to receive their awards. The organizers started with the project placed at position eight, which was the last one. My student and I just sat there waiting for our turn. We noticed that in all cases it was either a Vice Chancellor or project leader who read the acceptance speech. I decided to spring a surprise on my student and the entire audience. I was convinced that this was a rare opportune time to live true to our empowering approach that typified the "Amplifying Community Voices" project. My student would read the acceptance speech. I decided. I alerted him and indicated he should be ready to read the acceptance speech when our turn arrived. After feeble protestations, he agreed.

Now, on the big screen was the brief description of our project. We were invited to receive our prize. Our time had arrived. Note that we had not told our Vice Chancellor that we were to receive an award for third position! It was part of our strategy to surprise him. I glanced in his direction. He was seating with other Vice Chancellors from South Africa. That he had been swept off his feet was an understatement. As he leapt to his feet, camera in hand, to capture images of our special moment I could tell that we had accomplished one special mission—to change our institutional narrative on community engagement as a key pillar of university business. There he was transforming into a photographer and videographer as attention shifted towards us. My student delivered the speech so flawlessly, exhibiting immense passion that stunned the audience. I could hardly believe this was unfolding in my eyes and ears. Pride and ecstacy swept through my entire system.

What happened after the speech remains probably our most significant marketing stroke ever. Instead of the audience seeing me going to the podium to read the acceptance speech, my student had done it in a commanding over. Immediately after the speech, we received congratulatory hugs and handshakes without any restraint. Proceedings ground to a halt. We had done it! We coined it, 'positive commotion'. For two minutes, my student had simply confirmed what I knew already—that we were 'growing our own timber'. Here was a student in his final year of studies towards the Bachelor of Arts Youth in Development degree doing our university proud. The salvo of flashes of cameras that accompanied his speech told me that in that short space of time we had delivered the moment of the conference and vindicated the Talloires Network. This was to receive special mention when the Chairperson of the MacJannet Foundation delivered his remarks at the end of the ceremony. Those two minutes had placed the University of Venda firmly on the international map.

Video footage of the 'positive commotion' we caused in Madrid set a new tone for engaged work on our campus. Behaving like a young boy who has received his first favourite toy, our Vice Chancellor shared the video when he presented the MacJannet Prize of Global Citizenship to the Senate and other for a when he returned to our university campus. In May 2020, that event and other developments that unfolded in later years, our university adopted "A leading university in engaged scholarship" as its new vision. We glow in pride that this partly confirmed the power of our two minutes in Madrid in 2011.

Conclusion

As I wrote this piece, I felt the urge to draw conclusions from these stories. However, consideration of the diversity of contexts and experiences in development practice convinced me to shelve that thought. I leave that challenge to you. You are likely to agree with me that the experiential stories I have shared above are multi-faceted.

They carry lessons drawn from a wide range of national, cultural, policy and legislative environments, among others. Thus, it is only fair to let you draw your own conclusions. Where possible, compare your views with those of persons in your development networks.

Joseph Francis is a Full Professor and Director of the Institute for Rural Development at the University of Venda in South Africa. He has worked as a university academic for more than 23 years, during which time he has extensively engaged with grassroots communities in various countries in southern Africa including Botswana, Lesotho, Mozambique, South Africa, Swaziland and Zimbabwe. He holds a PhD in Agriculture from the University of Zimbabwe, specializing in integrated crop-livestock systems. As part of his academic and leadership roles in rural development, he is involved in research; Masters and PhD student supervision; mounting rural community-based development programmes; and establishing strategic partnerships for rural development.

Chapter 20
Early Career Experiences and Lessons in Rural and Urban Development

Joseph Kamuzhanje

Introduction

My journey in rural and urban development has exposed me to many cultures. I have come to appreciate the interconnectedness of the continent's culture despite the large diversity. The stories presented in this chapter document experiences, challenges and lessons derived in conducting research in rural Africa. During the past 30 years, I have experienced the highs and lows of development both at a personal and professional level. These experiences have steeled my resolve and commitment to appropriate the lessons to benefit future generations. I have been fortunate enough to work in various sectors including; Government, non-Governmental organisations and the private sector during my life in development work. This workspace places me in a good stead to have an all-round view of how rural and urban development can be used as a vehicle for effective and sustainable development. Five cases are presented here. The first case study is about the intricacies of access and mobility in rural areas. The second addresses the naivety of a fledgling development worker. While the third part highlights the need to always uphold professional standards, the fourth underscores the intricacies of doing development work in a different socio-politico-cultural environment. The last section of the stories underscores how political instability could hamper people's development.

Accessibility of and Mobility in Rural Areas

In 1989, I was in my second year at the University of Zimbabwe doing research on service provision at rural service centres in Gokwe District, Zimbabwe that was supported by the Swedish International Development Agency (SIDA). As part

J. Kamuzhanje (✉)
Coopers Zimbabwe Pvt Ltd, Harare, Zimbabwe

T. Madzivhandila et al. (eds.), *Development Practice in Eastern and Southern Africa*,
https://doi.org/10.1007/978-3-030-91131-7_20

of the project, I had to visit all the rural service centres in the district to carry out some assessments. I did the assessments ahead of schedule in all the centres except for Nembudziya in the north.

As I was planning to travel to Nembudziya, which is about 70 km from Gokwe Centre, I was advised that this mission would be an easy task. I woke up early in the morning to catch a Harare-bound bus that would drop me off at Kuwirirana where I would then find transport to Nembudziya. I decided to start walking towards Nembudziya after alighting the bus; it helped that I was super-fit from my playing football. After walking for less than a kilometre, a big truck came rattling by. Had I known, I would have at least tried to flag it down, but I was so convinced that it was one of the many vehicles that would be going in my direction. As it turned out, it would be the last! I walked the rest of the 26km in the searing heat of Gokwe. It took me the better part of 4 hours to complete the journey. I arrived at Nembudziya drenched in sweat, dusty, thirsty and hot. I was quite thirsty but there were no cold drinks and there was no water. All that was available was ice-cold beer. Although I am a teetotaller, I resolved that I had to take one bottle of beer otherwise I would suffer from serious dehydration. Fortunately, a truck with milk products arrived just in time and I did not have to drink the beer! I had no time to rest so I proceeded to conduct my assignment.

As soon as I was through with the work, I decided to look for transport to take me back to Kuwirirana so that I could then connect to Gokwe from where I had arrived earlier in the day. There was a bus coming from Karoi going straight to Gokwe, but it was a ramshackle vehicle and I decided that I would just get to Kuwirirana and then proceed on my journey using the bus that I had boarded in the morning and which was in good shape. However, I was later informed that my preferred bus to Gokwe had a breakdown and was not coming at the scheduled time, which should have been around 9 p.m. I had a choice to make. I could either sleep on the floor outside the shops or I could walk to Gokwe, which was 42 km. I chose the latter. This was not just a 42-km journey. It was a very difficult terrain and it was at night, too. I caught up with a guy who was going the same way but who only had to travel as far as 25 km. We started talking and that made the walking much better. As we approached his place, he asked what I was going to do. I told him I would just continue. He didn't think that this was the best of ideas because there had been several sightings of "spooks" in the area, which also explained why there were zebras which grazed with cattle in the area and no one dared touch them. I think he was just trying to make sure that I got some rest but there was no need to instill fear into me! He explained the issue by saying that there was a truck that would leave his place around 5am in the morning to go to a place 6km from Gokwe to collect river sand so I could jump on to that truck and then if I wanted, walk the last 6km home. If I missed the truck, I could then catch the early morning buses to Gokwe.

At 5am the following morning, the truck was surely there. But it was a big truck, a very big truck. I couldn't fathom myself getting on it. I was too "cool" to ride in such a big truck! So, the truck left, and I waited for the buses. By the time the buses started coming, the place was already full of people intending to travel as well. However, the preference of the "bus conductors" was for people travelling straight to Gweru!

With no option left, I started walking the 18km journey that remained. This story highlights some of the transportation challenges faced when working in rural areas of Africa. In addition, it teaches us not to be choosy when presented with limited opportunities.

Setting Off on the Wrong Side of the Compass

I joined the Department of Physical Planning in the Ministry of Local Government in May 1992 immediately after graduating with a BSc in Rural and Urban Planning from the University of Zimbabwe. I was posted to Masvingo Province where I was the youngest of the planners and technicians. I was excited, quite ready and enthusiastic to serve in the noble profession and change the world.

My first field assignment came about almost by accident. One of the NGOs wanted to set up a refugee reception centre at the Sango Border post in Chiredzi District. The planner who was responsible for the district asked me to go on his behalf as he was not available for this assignment. I had never been to Chiredzi and I never interacted with NGOs in the past. I was still on my probation and had not yet officially represented the ministry and department before. I knew this was a perfect opportunity to set off my planning profession.

The drive to Chiredzi was quite exhilarating for me. The change in climate, the physical environment, and settlement patterns as well as the sugar cane fields in Triangle and Hippo Valley were nothing I had seen before. However, the respect that I got from the support staff from the NGO that I was travelling with was more exciting for me. I was granted whatever I requested. I never imagined that I could wield so much power!

At Sango Border Post, we attended a security briefing and were advised that we had to carry out all our work in the car due to the dangers of landmines. This period was during the war between RENAMO and the Mozambican army and landmines had been planted haphazardly to curtail the movement of civilians running away from the war. Whilst the Zimbabwean army had done all it could to clear the landmines, we could not take any chances.

As the lead, I was now supposed to show the team where the reception centre would be sited. I took out the site plan, laid it on the table and then indicated to the team where construction work should be situated. After that, we went out in the car for the physical inspection. I pointed to the area where refugees from Mozambique would be received by the officials of the Zimbabwean Government. Everyone in the team was satisfied that we had chosen the most appropriate site and it was a job well done.

I went to debrief the responsible planner on a Monday following the site visit. The whole visit was still quite fresh in my mind and as I looked at the site plan again, it suddenly dawned on me that I had held the site plan the wrong way when we first visited the site. “My” north was in the south and the reception centre would be virtually built in the opposite direction. I could not tell the officer what had happened. I told myself that I would volunteer to go again to put the pegs on the ground.

Luckily, a month later, there was a ceasefire in Mozambique and the war had ended; there were no more refugees crossing into Zimbabwe and the reception centre did not have to be built.

I am very happy that this incident happened very early in my career as a planner and that no lasting damage was done. Afterwards, however, I always made sure that my compass was always right. But more importantly, the incident also taught me to be extra careful in everything that I did. A similar mistake in the future might not necessarily have a good ending!

Principles Matter

One of the most contentious issues that physical planners must deal with is "illegal development". According to the Regional, Town and Country Planning Act, a local authority has the powers to demolish any structure that is deemed to be illegal. That is the case that confronted me when I joined the Department of Physical Planning in 1992 in Masvingo Province and was assigned Bikita as my district of responsibility.

In 1991 and 1992, Zimbabwe suffered one of its worst droughts in living memory. The Grain Marketing Board (GMB) depot at Nyika Growth acted as a point for people to buy maize grain, after which they had to find milling facilities to process the grain into mealie meal. To mitigate this challenge, a few business people set up a milling company and decided that the factory should be as close as possible to the GMB so that people would not have to move long distances to access the facility. In fact, the company would just buy grain from the GMB, process it and the people would come and buy the mealie meal from them.

It was only after the structure was put up that trouble started. The milling company had been established on land belonging to the Ministry of Public Service and within its training centre and was therefore, in planning terms, illegal. This issue was compounded by factionalism within the ruling party, ZANU (PF). An issue that could have been resolved by referring to the legal provisions suddenly became very political and a battleground for the two factions in the province. For illustrative purposes, the factions herein are referred to as "X" and "Y" with faction X being in favour of declaring the milling company structure illegal and Y against the idea.

The issue was tabled in the Planning and Works Committee which was dominated by faction X and a recommendation was made that the structure should not only be demolished but that the company should meet the costs of the demolition. The issue was then tabled in the Full Council which was chaired by a member of the Y faction. There was deadlock after deadlock. Whenever the department was invited to the meetings, my boss, the Provincial Planning Officer would accompany me. However, when the issue came up for discussion, he would find an excuse to sneak out so that he could "carry out some other business". After all, I was the planner for the district and so should be able to talk on behalf of the department. So, there I was, still green behind the ears in terms of planning and defending a position on one of the most controversial issues ever in the province. All that I had was the Regional, Town and Country Planning Act and I stuck to it like a leach.

I was being bombarded from all angles, including from the chairperson of the Planning and Works Committee who belonged to faction Y but had been outnumbered by members from faction X. Even then, when he came to the Full Council, he was supposed to support his committee's decision. In contrast the chairperson made a statement that has stuck with me ever since. Looking pointedly at me in one of the Full Council meetings, he noted that "a plan was just a piece of paper. We can tear it up and draw up another one if it does not address our interests!".

Almost 30 years later, the structure is still standing. There is a Council resolution to demolish the structure, but it was never implemented. I have always told myself that the Bikita incident defined the development practitioner that I have become. It taught me to be principled, to be strong and to stand up for what I believe in. It taught me that as long as the issues were clear, no-one, [and not even politicians] could stand in my way. Although the issue was never resolved, the experience developed my presentation and articulative skills.

Facilitating Training Workshops

In 2011, I was part of a team that was contracted by the South Sudan Government to train its Local Government Administrative Officers (LGAO) on their roles, responsibilities and guiding legislation. I was posted to Bentiu in the former Unity State. You must understand that it was still the time when South Sudan had just gained independence and so things were not exactly in order. My first experience of Juba International Airport was chaotic. When I landed from Harare, it was very hot, there was one arrival and departure lounge, the room was small, and I could not speak or understand the local language! By the time I made it out of the airport, I was so exhausted and all I wanted was to find a place with a shower and then sleep. The trip to Bentiu was on a Saturday and we flew in a small 6-seater plane, which I thought was still gaining altitude by the time the pilot was preparing to land. From the sky, I could see a gravel runway and there was a herd of cattle grazing. So, the pilot had to circle the runway as the "airport" officials chased the animals away.

I went to the venue of the training on Monday morning. It was just a big, open room. It had space for 30 people but there were 50 people waiting outside. There was no electricity and that meant no air-conditioning. It was 9 a.m. but I was already sweating profusely. Then the magnitude of the task hit me. I was the only trainer and I had to deliver the training to this crowd whose ages ranged from 22 to 70. In addition to my challenges was the language barrier.

In 1983 when Jaafar Nimeiri became the President of Sudan, not only did he introduce Sharia law, he also issued a decree that Arabic was the official language and that all Government communication was to be made in that language. This was not a problem for the north as it was mainly Muslim but a big challenge for the south which was predominantly Christian.

In the end, I discerned that about 5% of the participants could communicate well in English, 80% spoke and wrote English but as a second language and then 15% could not understand English at all. This was not the only problem. Around 50% of the trainees (both personal and professional) had never attended a training workshop before and the participants were split in half along the two main tribes, Dinka and Nuere. Tribalism is a very real problem in South Sudan and failure and success of any development project depends on how tribal politics play out.

To many, this training was the first time that they saw the documents that were supposed to guide them in their work, such as the Local Government Act or the Constitution of the Republic of South Sudan. The training itself was a process. First, there was a lesson in English where I would have to explain and describe every word that I used, and then find an appropriate example to drive the point home. Besides, I would use the word in the legal sense and then explain it as used in the document. I would then have to get someone to interpret in Arabic before going to the next sentence. And this would happen every day and in every session of the training. It was the same thing during the weekly assessment tests. I used the open exam system where the trainees were free to refer to their training materials for answers. I still had to explain every question and almost gave them the answers in the process!

Before I left South Sudan, there was an end of training get-together party that I will never forget. I was not prepared for the outpouring of gratitude and thanks that came from each participant. The trainees contributed money from their allowances to buy me traditional South Sudanese artefacts, which I have kept until now. Every time I look at them, I remember the 70-year-old man who attended every training and in his broken English, contributed to every discussion and was the calming influence whenever tribal issues threatened to flare up. I also remember how this experience humbled me and taught me to be patient and never to take anything for granted.

That Is not The Way to Hold a Gun!

In 2017, I visited South Sudan for the third time. I went there as part of a team that was carrying out performance assessments for the counties (districts) that were benefitting from a World Bank project. I had fallen in love with the country on my first visit in 2011. Perhaps it was because I was caught up in the euphoria of a newly independent nation. I really wanted the country to succeed and set an example for Africa. For a country that is blessed with so much potential, I felt it was a possibility that countries could move from strife to prosperity aided by its own internal resources. By the time I came back in 2012, the situation had changed dramatically. The country was on the verge of a war with its neighbour, Sudan, the economy had begun to shrink and there were rumours about an impending civil war between the loyalist armies of the President and those of the Vice-President.

In 2011 and 2012, one of the things that had amazed me was just the number of guns that I saw everywhere where I turned. This was a very new experience to me. So, when I got back in 2017, I somehow understood that after the civil war, there would be security issues especially outside of Juba, the capital. Even with this background knowledge, I was not fully prepared for what I came across. On my first trip to the field, I saw a man riding a motorcycle with an AK-47 slung on his back. There was a man with a gun buying beer at the shopping centre. It seemed to me that there were guns everywhere.

On one occasion, we were visiting a school that was being constructed by the community with support from the project we were assessing. Part of the school project included a vegetable garden that had a water pump. As part of the assignment, we went to visit the garden and I was so caught up in what I was doing, I did not notice the two young men coming into the garden. When I looked up, I could not believe what I saw. These were two young men, barely 15 years old, raggedly dressed with no shoes but each of them holding an AK 47 as if they were holding a cattle whip!

In my excitement and confusion, I asked the team that I was travelling with whether I could have a photo with the two boys. I was told that the boys could be temperamental, but they would ask them. There was a short exchange of words in the local language, and I was asked to pose with them for the photos. I stood with the boys and then held the AK 47.

This was my first time to hold a gun, let alone come so close to gun-wielding people. After the photo session, I was told that the guns were loaded with live ammunition and that by that age, the boys were top marksmen in terms of their dexterity in gun-wielding.

Later that afternoon, we saw the same two guys, only that this time, the other boy did not have a gun. I asked what could have happened and I was told that he could have had sold the gun. I left South Sudan thinking of what the country was going to do to move away from such a dangerous situation. The experience offers a thorough reflection on the importance of peace in the development process.

Conclusion

The stories shared in the chapter cover different aspects of the development activities in which I have been involved over many years. They indeed represent my experiences in various assignments that I have navigated throughout my career. While they are very personal, they may have reflected, in some way, other development practitioners' challenges, opportunities, and experiences. One of the things that I have learnt over the years is that development practitioners do not learn about development by accident. The experiences acquired in development work seem to occur naturally. The fact that the learning is not only existential but also practical means that we need to have an open mind on how the development process should evolve, how we should be involved and how we interpret development. While I have been fortunate to traverse this development path, I am under no illusion that I have achieved the goal; there are still many more to be accomplished in years to come. Past and future experiences can always serve to make me a better human being in the end.

Joseph Kamuzhanje Is a development planner with over 30 years of experience in rural and urban development. Over these years, he has experienced the highs and lows of development both at a personal and professional level. These experiences have steeled his resolve and commitment to spend his life converting them into lessons for future generations. During his career he has been fortunate enough to work in Government, non-Governmental organisations and the private sector. This places him in good stead to have an all-round view of how rural and urban development can be used as a vehicle for effective and sustainable development

Chapter 21
Experiences of a Young Academic Development Practitioner

Pertina Nyamukondiwa

Introduction

I am passionate about research, facilitation and co-development of tangible community-driven solutions for existing challenges. While in the field, some experiences were frustrating while others were encouraging. But in both positive and negative experiences, there are important lessons for practitioners, researchers, and organisations in any field of development.

I was introduced to developmental research while I was doing field research for my masters' degree in one of the villages about 35 km outside the university. My study focused on resilience of rural communities to the effects of climate change. The study followed a mixed-method design. I had finished collecting qualitative data in the primary village called Mabayeni. The next stage was to use results of the first phase to do a quantitative confirmatory study in nearby villages, which were in the same ward. The ward had nine villages. The qualitative phase had gone very well, so I was very excited. I asked the village headman from Mabayeni village to connect me with other headmen from the remaining villages. He gave me their cell phone numbers. I made appointments with each one of them. I met them and explained what the research study was about, its purpose, objectives as well as the kind of information I was looking for.

I requested for names and contact details of male and female youth who would serve as research assistants. I got their contact details and set up a meeting with them. In the following week, I met with a group of six youth members. I explained the purpose of the research, its objectives and the roles of the group members who would serve as research assistants. Moreover, I explained the non-financial benefits of the study to the group. I also took them through the data collection tool to ensure

---

P. Nyamukondiwa (✉)
Adjunct Senior Lecturer, PhD in Rural Development, Masters in Rural Development, Honours in International Relations, University of Venda, Thohoyandou, South Africa
e-mail: Pertina.Nyamukondiwa@univen.ac.za

T. Madzivhandila et al. (eds.), *Development Practice in Eastern and Southern Africa*,
https://doi.org/10.1007/978-3-030-91131-7_21

that they understood all the research questions. Everything went well—or at least so I thought.

I then went to the village headmen to give him feedback and to agree on a date that I would collect data. In fact, I requested him to organize a meeting on my behalf, with residents on a day convenient for them. I would then come and distribute questionnaires so that they complete them in my presence. He indicated that he would announce at a monthly community meeting to obtain a suitable date from community members. The monthly meeting was due to take place in a week's time. This was good! So, I waited. He then gave me a call after that week to give me the date and venue, which had been agreed upon during the community meeting. The meeting was to take place on a Saturday, at a community hall. The starting time was 9 a.m. I was very thrilled that all the plans were coming together. I called my research assistants to remind them of the date. They were all ready.

I arrived at the venue at 8:10 a.m. on that day. I was supposed to meet with the research assistants at 8:30 a.m. for last minute preparations and allocating questionnaires. At 8:30 a.m., the research assistants had not arrived. I waited. Two of them eventually arrived at about 9:30 a.m. We waited for another hour until some people started arriving. At around 10:40 a.m., with only two research assistants and huge group of youth, we started with the meeting. The group was big, about 40 youth male and female combined. There were no adults or elderly people. The absence of other groups worried me. But I took comfort in the large numbers of youth. From my experience in community engagement, it was very unusual to get those large numbers of youth participating in research. It was a bit strange too! But I was comforted, nonetheless.

As I explained the purpose of the meeting, some male youth at the back of the hall started making noise. When I enquired, they indicated that they had not come to do research. "We were told you were bringing us jobs! We are here for jobs not your research", they shouted. I tried explaining but it did not help. Most of them walked out in protest. Only five female youth stayed and were still keen to complete our questionnaires. They confirmed that they had been told they were coming to get employment but could not divulge exactly who had told them. The huge group had come from two villages, not one as had been planned. But what had gone wrong? Was it because I did not promise financial benefits? At what point did the communication break down? Unfortunately, these questions remained unanswered.

Time was ticking. I was compelled to change the sampling technique. I went back to report to the headman who also could not explain what had happened. I asked for permission to go and administer questionnaires door to door. The headman promised me that he would notify community members of my visit in advance. Luckily, the five ladies who remained agreed to assist me with collecting more data from the remaining villages. What a bad start to Phase 2 of data collection! Thankfully, we managed to collect data from all the remaining villages. What I learnt from this experience was that investing time in social preparation is very crucial before undertaking research in the field. As researchers, we often do not take time to learn the socio-economic or political dynamics within our target areas of study. The reason could be because our research projects are mostly time-bound—we just want to collect data and generate knowledge within the confines of our deadlines, or those of our funders. In my case, more time was indeed wasted when I had to change the sampling technique and visit community members door-to-door. If I had spent more time interacting with community members, I would have understood the dynamics and possibly foreseen and averted any potential risks. Yet up to now, I still do not understand what really happened.

What Have You Brought Us?

This was my last day of data collection in Mabayeni village. I was concluding the first phase of data collection, which was qualitative and exploratory in nature. I had had conversations with different community groups including children as young as seven years old. I interviewed groups of children, youth, adults and the elderly. I needed to get their understanding of climate change including its effects in the village. This was part of my master's degree. I had been in the village for about three weeks. On this day, I interviewed a group of elderly women. This was one of the most exciting groups to engage because of their sincerity and willingness not just to respond to questions but to also educate. This is a group, which seemed to have all the time in the world. So, I set aside a day to sit with them. The mood was quite relaxed. The fact that they had seen me in the village several times engaging other groups made it easy for us to jell together.

They shared their observations of climate change in the village and went into depth explaining traditional practices, which used to take place in the village. Towards the end of the engagement, I expressed my appreciation of their time and contributions. On a lighter note, one of them invited me to become a daughter-in-law in the village. They all laughed. The mood was still good even after spending the whole day with them. I was happy. Then one of the women who looked to be the oldest asked me, "My child, you see this part of the district is very dry, what have you brought us from Thohoyandou? You know in Thohoyandou there are various fruits. There are Bananas, Oranges, Naartjies, etc. What have you brought for us?" I did not understand. In fact, I was confused. At the beginning of the engagement, I had explained the nature of the study—that it was an academic project, I was a student, I had no budget—I thought. "Fruits?", I asked.

Another lady, who looked the youngest came to my rescue. "You know what, many students have come to this village to conduct their research for purposes of studying as you are saying. We spent time with you giving you information and you go and graduate. Granny is asking how she will benefit from the study? How will the community benefit because we are tired of helping you to graduate yet our problems don't change. That's what she is asking?" I was dumbfounded. I never expected that. I did not know what to say. So, I simply repeated what I had said in the beginning although I knew this was not enough. I was challenged. Surely it was not right to just talk about problems and write research theses and leave. If all the research that we did went beyond investigating and evaluating challenges to developing tangible solutions, the world would be a much better place. My mind was troubled. I got a wakeup call.

As I walked out of the village in the scotching heat, I saw long snaking queues of people waiting to fetch water at a community tap. Water was a challenge in the village. In fact, water scarcity had emerged as the most pressing effect of climate change. I had to do more than just talking about it. This experience inspired my PhD research project, which focused on co-developing prototypes of a technology for harvesting rainwater for domestic use with community members. I took the research work further. My Postdoctoral research study focused on testing the technology, which community members preferred out of the many ideas that came out. I received funding from the Royal Academy of Engineering to finish the testing process (proof of concept) before the technology can be implemented. The hope is that it serves as a tangible solution to water challenges in the village. A fruit that I am confident the community will enjoy. Very tough questions I was asked! Yet they inspired transformation in the way I now see research; I now maintain a deliberate bias towards producing tangible solutions.

We Just Want Beef

I was collecting data for my PhD research project. The project focused on developing a community-informed technology for harvesting rainwater for domestic use. I had had several community entry meetings with community leaders. They understood the project's purpose and objectives. A community-based research team was constituted. Community leaders successfully mobilized other participants as promised and availed themselves for research discussions. I took some time to educate participants on what the research was about and the potential benefits for all stakeholders. Successful completion of the project was going to help communities with provision of a scarce resource—water. I expected community members to be excited about the research project. However, in the group discussions, as I walked around, I noticed that some participants were not up to the task. Instead, they were just doing it for compliance. One member of the male youth group openly confessed that they had only come to the engagement to get food. "We are here for the beef my sister; we just want meat. We don't believe in this research you are doing", he said.

As I walked around, I noticed that a group of adult women did not believe that the study was going to bear any fruits although they were answering questions honestly. "We want water here; we are tired of these meetings? Are you sure you are going to help us get water?", one woman asked. I learnt that it was not because community members had bad attitude towards me or the research. It was because they had seen many researchers who collected data and left (many without even giving feedback on findings), and politicians who left empty promises. They had lost hope in getting help from outsiders. It was only after about two months or so that I noticed real interest in the research. This is the challenge that researchers, politicians and other groups have created. The question is how do we rectify this? What do we need to do to regain communities' trust? My view is that we all need to become responsible, ethical, solution-driven practitioners who live up to their promises.

Resilience in Developments Practice

I had been to the village for a while. I started off the journey and picked my research assistants on my way to Mabayeni village, which was my study area. I had six research assistants on this day. The distance from campus to the village was about 45 km. Barely 20 km into the journey, our car broke down. The engine had overheated. We tried pouring water into the engine, but it did not help much. We tried calling for assistance from friends and colleagues. We were not lucky.

We called community leaders in the village to notify them about the delay. It was shortly before 9 am and we were still stuck by the roadside. Community members started calling to find out if we were coming. We thought of cancelling the meeting, but it was too late. The catering lady had already bought food and cooked most of it. We would have to pay for the food anyway. We then explained our situation to the catering lady and asked if we could use her car. She agreed. But she indicated that one of us would have to go fetch the car because she was busy cooking. She stayed in the neighbouring Jilongo village. I then sent one of the research assistants to the village to fetch the caterer's car while we took our car for fixing. Word! The taxi took long to fill up! It took the research assistant about an hour and half to fetch the car. He finally picked us from the mechanic and we headed for the village. We arrived at the venue shortly after 11 am. All the community members had left except for two leaders. Despite updating them on what was happening; we had taken too long.

We asked the leaders if we could still call community members to let them know that we had arrived. We used a combination of phone calls and physically visiting nearby houses to invite them back for the engagement. Fortunately, they understood and came. We started late, shortly before 12 mid-day, but we had one of the biggest turn-out. No resources were wasted. Despite the challenges, we managed to have our meeting. The lesson I learnt on the day was that no matter the challenges, resilience pays off. We had a very bad start, but it ended up as one of our best days in the field. We had also created a good relationship with our local caterer. She assisted us with her car at no cost. A display of 'Ubuntu' (humanness). We were exhausted, but we left the village smiling.

The Wrong Data Collection Tool

After collecting data, I had to do part of the analysis in the field with research participants. I had to use participatory rural appraisal techniques such as seasonal diagramming, matrix scoring, participatory mapping among others. On this day, I was doing a combination of seasonal diagramming and matrix scoring. The objective was to find out at what point during the season community members experienced water scarcity the most. I had bought bean seeds for participants to use when carrying out group matrix scoring. After the exercise, I humbly requested participants to return the seeds. Regrettably, it was too late. The 2 kg packet of beans had disappeared. I didn't worry much about it. Instead, I figured I had not properly communicated that I needed to use the bean seeds in future engagements.

Thus, in the next engagement, I made sure to make the request much earlier. Soon after giving instructions of the research task, I requested all participants to return the bean seeds back after the ranking exercise. Still, nothing was returned. When I investigated why members were not keen to heed the request, I was told plainly that I was playing with food. That's when I realized that my data collection tool was inappropriate. Had I used corn seeds, would I have come across the same challenge?—I wondered. Possibly, what made my choice of research tool worse was that it was considered relish. Perhaps, stones would have been a better option. Of course, these were some of the lessons I learnt after taking a moment of reflection.

The Mighty Marker

We had just finished with our research tasks for the day. We had conducted focus group interviews and done matrix scoring and seasonal diagramming. Each group had a set of flip charts and two markers to use. After the discussions, I requested groups to return the flip charts and markers. All the groups returned their materials except for one group—the female youth group. They returned only their flipcharts. When I asked for the markers, the group members pointed at one young lady, who looked slightly younger than me. So, I smiled and turned to her. I gently asked for the markers. I explained that I needed to use them in future engagements. The young lady plainly refused and indicated that she was going to keep them for her 3-year-old son to use. I added that if she needed them, I would give her on the last day of data collection. She was unperturbed. I thought she was joking. She maintained her stance and explained that she had been coming to the meetings for some time. She was rewarding herself with the markers. I found it difficult to believe.

The same day, one woman from the adult women group had approached me and requested for a marker during group discussions. She explained that she needed it for their "Stokvel"—a maize savings club. She was the secretary of the group and, therefore, responsible for recording maize contributions from Stokvel members. She needed to mark the maize sacks as she received them. I understood their great initiative and promised to give her two markers at the end of the engagement—which I did. Reflecting on the two incidences, I was left with some questions. Did the young lady take the markers just because she was being disrespectful or because this was a manifestation of frustration, which had built overtime? Perhaps, she was also annoyed by outsiders; researchers and politicians who visited the village, spend time talking about problems, and made empty promises (just as the other groups had expressed on different occasions). Or was it because she felt she could do it because we were about the same age? Did she really need the markers for her child? Why didn't she ask? These unanswered questions strengthened my conviction that development practitioners need to improve their conduct when working with communities. This will reduce unnecessary conflict and tension, which hinder development progress.

The Discovery of the "Swidhakwa Group"

The research task for respondents on this day was to design low-fidelity prototypes in their groups. Seven groups were expected to each design two prototypes (one thatch and one zinc-based). Community members were divided into seven groups namely children (boys separated from girls), youth (male separated from female), adults (men separated from women) and the elderly. Men had not turned up on the first day for the prototyping task. This was the second day of the task and the same group was missing. A large group of men was seen drinking traditional beer at one household. On realising a possible opportunity, a short meeting was held with research assistants to explore the possibility of inviting the men for prototyping. After getting a positive response, another meeting was held with a few community leaders about the same issue. Nonetheless, community leaders had some reservations. This was a group of drunk men who could potentially disturb other groups. It was finally agreed that "the drunkards" could be invited if they would be monitored to ensure that they don't become disruptive.

On invitation, the men agreed to participate in the prototyping exercises on one condition; they wanted to continue enjoying their drink. This was accepted. They brought their bucket of alcohol and made prototypes while drinking. They did this under a tree, away from the other groups. It was evident from their conversations that the men were excited to participate. They wanted to prove a point to the rest of the groups. "They look down upon us. We know that they don't take us seriously because we are drunk. We will show them. This is our thing," they were heard saying. They even proudly named their group "Swidakwa" meaning drunkards. This group of men worked together very well. They did not cause any problems and finished constructing their two prototypes in one session. The other groups took two to three sessions to complete.

During the evaluation session, one of the men's prototypes were selected as the best by all groups. This was amazing because other participants did not expect "the drunkards to produce meaningful prototypes." In as much as I found the experience exciting, there were inevitably some ethical questions which arose. Was it ethical to involve men that were intoxicated? Well, I was faced with two choices; going ahead without the group of men (which would possibly be a bigger challenge because I would be missing a critical perspective), or involving them under close monitoring to ensure that they would not cause harm to other participants. This is one example of a situation, which challenges researchers and practitioners to be innovative while at the same time ensuring validity and reliability of research results.

The Inopportune Engagement

The agreement was to start engagements at 7 a.m. on the day since it was a National Youth Day holiday. The research team arrived at 6:45 a.m. It was not until 8 am that a few community members started arriving. By 9 a.m., only seven people had arrived. The engagement only started at 10:30 a.m. with 30 participants. Others arrived very late. The turnout was terrible. Out of the expected 120, only 49 participants turned up on the day. Refreshments had already been organised and last-minute cancellations were impossible. It was a very cold winter day. Were participants discouraged by the weather? Was it the holiday, or was it something else? We could not help but wonder. We had agreed to start earlier than our normal 9 a.m. meeting time to give people time to go for Youth Day celebrations after the engagement. We had reminded our community-based research team to mobilize as we had planned. We had done everything we could. We could not figure out what had gone wrong.

We devoted some time to understand what had gone wrong before commencing with the day's programme. It emerged from the discussion that one of the community-based research team members had discouraged community members from participating because "there was no proper food but only bread". I had communicated with this team member that due to the large numbers of people that were expected, only refreshments and not the usual standard lunch would be provided due to financial constraints. She then used this information to discourage people from participating in the engagement. As a result, the feeling among the community members was that the researchers "were using people and saving money for their enjoyment at the expense of community members". It was very disheartening to learn that the person whose responsibility was to mobilize participants had on the contrary, discouraged them to participate. We felt betrayed. After the discussion session, I (as the project leader) apologized for the misunderstanding and took some time to clarify some issues related to budget, purpose of the research and ways in which the researchers and community would benefit. Then, the engagement proceeded with the few people that were available. Going forward, we ensured that we would explain the purpose, objectives, outcomes and potential benefits of the project to all participants in every engagement.

The Leader Who Failed to Deliver

The prototypes that were developed in Mabayeni village were supposed to be evaluated by the designers (community members) together with members from the external community who had not participated in the design process. A neighbouring village was selected to participate in this evaluation process because it was also experiencing severe water shortages. A meeting was held with the village headman to explain the background and purpose of the study. Permission was also sought to engage community members from this village. About 35 participants (at least five per interest group) were requested to join the prototype evaluation session, which was taking place in Mabayeni village. The headman of the neighbouring village was very supportive and indicated that he would call a meeting and make sure that the required representatives participated.

Three days before the engagement, reminder calls were made to the headman who indicated that he had already mobilised the people and relevant participants had expressed willingness to participate in the evaluation. On the day of the engagement, participants from this village did not turn up. Upon making follow-up calls, the headman indicated that the participants were on their way. Towards the end of the engagement, only four elderly women arrived. They indicated that they had received communication about the engagement the same morning. Efforts to establish what had happened from the headman after the event were fruitless. As disappointing as this experience was, the lesson learnt was that relying on one person when organising engagements is not enough. I could have spoken to more community members to make sure that the required number of participants was mobilised. Working with key informants could have helped.

You Should Have a Budget…

I was involved in a community development programme in one of the villages around the University of Venda. The community was made up of three sub-villages. Before the commencement of community activities, preparatory meetings were held. In fact, the community had reached out to the university's Institute for Rural Development (IRD) for help to develop their community development plan. The IRD promised to provide the requested support but indicated that its resources were limited. The institute would provide staff members and postgraduate students to do the work. It would also provide transport for the university staff and students. Volunteer students were recruited and trained. Dates were set. Community leaders indicated that they had sufficiently mobilized their people, and community members were ready for the engagements.

I was given the responsibility to coordinate the programme. The plan was to implement the Asset-Based Community Development (ABCD) approach in the community. This meant that we had to train "foot soldiers" from the community who would assist in collecting data on the different types of assets found in their community. In the first meeting, we introduced the programme. The turnout was impressive. Many community members especially the youth turned up. In the second meeting, the numbers reduced. We learnt from this meeting that many youth members were under the impression that they were coming to get employed. We took time to clarify the nature of the programme as well as their expected roles. In the same meeting, we introduced the ABCD approach and explained why it was an important tool for facilitating sustainable community development.

In the third meeting, the turnout had not improved. But we were comforted by the presence of the Ward Councilor. As a leader, we were hoping to request her to assist us in mobilizing the people. However, we were shocked to hear her complain about the absence of food as a reason for bad turn out. "You cannot organize a meeting without a budget. You must have a budget", she said. We could not believe our ears. Community leaders had sought assistance from the university, and we showed up. We did not know that we were expected to provide food as well. This is when we realised that the community was not ready for engagement. Thus, we held off all community engagements for a while and allowed internal engagements to take place. These would enable community leaders and members to understand the programme and the nature of IRD's involvement.

It appeared that whenever they needed the village youth to participate in community engagements, community leaders often used the employment card. Perhaps community leaders could not deal with the challenge of failing to convince this group to participate in development-related engagements and then decided to take the easy route—deceit. But to what end? Unmet expectations on the part of the youth resulted in reduced turn out for the rest of the engagements, which compromised the progress of the programme. As for the Ward Councilor, what compelled her to make a U-turn on us after being part of the group of leaders who approached the university for assistance? Did she succumb to pressure from community members? These questions remained unanswered in our heads as we waited for the conclusion of the community's internal engagements. Over and above, we learnt the importance of good leadership in community development. It is very dangerous to have leaders who are easily influenced by the masses as situations that are easily manageable can descend into chaos.

Conclusion

The above set of stories reveal several lessons for researchers and development practitioners. The trenches of development often demand practitioners to expect the unexpected and always think on their feet. There are always situations and circumstances that arise when carrying out fieldwork, which demand skills and attributes such as strong communication skills, being innovative, improvising, negotiation, resilience, and in some cases, a lot of patience. In addition, ample time needs to be set aside specifically for social preparation processes. This builds unity and destroys the "us" versus "them" mentality between practitioners and communities. Social preparation is a decisive factor in either the success or failure of community engagement projects and programmes. It is not only community leaders that must be aware of any development programmes taking place in their area; ordinary community members should as well. Investing in social preparation creates an opportunity for mutual understanding on socio-economic and political issues, underlying frustrations, and expectations from all stakeholders. Moreover, this process puts all community stakeholders at ease and prepares them for genuine participation. Most of the challenges I faced were as a result of not investing enough time in understanding the communities I was working with. I believe these challenges could have been avoided. Another huge lesson was that, in as much as it is important to undertake research and contribute to the body of scientific knowledge, there are real societal problems whose solutions can be developed through research. Thus, it is crucial for researchers to also think about developing practical solutions to problems, which our communities are facing daily.

Conclusion

Lived experiences of development practitioners in eastern and southern Africa: A Synthesis

By Joseph Francis, Sepo Hachigonta, Joseph Kamuzhanje, Tshilidzi Madzivhandila, Oluwatoyin Dare Kolawole and Shirley DeWolf

The primary aim of this book is to document diverse experiences of development practitioners in eastern and southern Africa. The compilation of these experiences primarily serves to provide teaching materials in colleges and universities, in addition to creating a veritable avenue for policy development. The stories, therefore, constitute a reflection of individual authors' perspectives about how they viewed development practice through their own lenses. The chapters are clustered under four sections based on their commonalities and the ideas they convey to the reader. While some of the stories accounted for the experiences of researchers on how they navigated their ways through seemingly difficult and precarious terrains in the process of conducting development-orientated studies, others provided a vivid exposition of practitioners' accounts from the frontiers of development practice including the vulnerabilities that female researchers experience during fieldwork. The book also illustrates how cultural dynamics, skills development and communication might impact on social change agenda and initiatives in certain places. The last part of the book is a compilation of some reflections on how practitioners navigated social and political terrains as they engaged communities when implementing development initiatives.

While development projects have positively contributed to the livelihoods of grassroots communities, numerous negative perceptions have been created on their actual value. An example of such perceptions is the notion that development practitioners and academics would engage in self-gratification and personal gain while conducting research and implementing projects in rural communities. They do so without necessarily reflecting on how their actions might impact on grassroots communities long after the conclusion of their assignments. This is a recurrent issue in local communities where people feel development workers abandon them immediately after achieving their professional aspirations.

T. Madzivhandila et al. (eds.), *Development Practice in Eastern and Southern Africa*, https://doi.org/10.1007/978-3-030-91131-7

Culture dynamics are shared when working in multicultural teams and there are differentiated viewpoints of local communities between African researchers and their European counterparts. A story showcasing how female development practitioners navigated the thin line between conducting research and observing community values as against the experiences of their European female counterparts is a classic example of these dynamics. Also, the need for a sound ethical conduct in the development profession is emphasised throughout the book. Without necessarily compromising work ethos, development practitioners' skills and innovative strategies are underscored in the book as a negotiation tool that might prove handy when dealing with grassroots communities. In addition to this, the book highlights how some innovative communication tools can be used to advance development work. The use of Theatre for Policy Advocacy (TPA), for instance, finds relevance in promoting community participation and involvement in evidence-based policy advocacy in broad socio-economic issues affecting people's livelihoods and development. Overall, most of the chapters converged in many places within the context of the challenges that field-based, development practitioners faced.

Challenges and Opportunities in Development Practice

Challenges and opportunities in any human endeavour are not mutually exclusive. In development work, some challenges might be ecological, social and politico-cultural in nature. However, issues constituting a challenge in a geographic and socio-politico-cultural setting might present opportunities for growth and development if they are objectively analysed and adequately addressed to achieve a desired objective.. This book outlines some challenges encountered by researchers working in dangerous wildlife areas. It also underscores the key barriers, which women in rural communities faced, and provides examples of how they could be empowered to share their perspectives on development projects especially in scenarios where the support and participation of their male counterparts are genuinely sought and copiously included in development interventions. While much effort is needed to increase women's leadership roles in the economy and polity, the book sheds positive light on how their roles are beginning to shift within the development process. Although differences in cultural identities might engender some constraints in following through with some development objectives, they might achieve the desired goals if properly harnessed. Considering the unique cultural diversities of African societies, the book explains some communication barriers and challenges encountered when implementing projects. The use of English and French in sharing information on development projects within a local context provides a good example of how social interactions and communication of development project ideas might be constrained especially in rural areas.

Many other salient challenges that development experts encounter include poor road infrastructure in remote areas, harsh environmental conditions and the social stigma associated with working in culturally diverse locales especially where a public health solution is needed to either curtail or contain disease outbreaks such as Ebola.

Policy Issues

Policies are instruments or guidelines used in driving development objectives. Formulating and implementing appropriate and quality, pro-poor policies are likely to impact significantly on progress and development of a community. For any development work to genuinely appeal to intended beneficiaries, clear short and long-term objectives are needed. The abilities of elite policymakers to understand any existing problem and their willingness to discard personal interests or gains to realise the public good might be the trigger for achieving progress. Thus, policymakers have various options for managing development problems, 'including at times the option of not addressing them' (Grindle & Thomas, 1991, p. 2).

That communities perceive development experts as grandstanding and selfish in their conducts cannot be ignored. This highlights the need for development experts to rethink their professional conduct and strategies to change the negative perceptions and attitudes of communities towards them. It is one thing to consult or engage communities in a development research or project. It is another thing to always sustain engaged interactions with the communities and provide adequate feedback whenever it is needed. Therefore, development practitioners should continuously embed themselves within grassroots communities where they worked long after the completion of their primary assignments. This is a sure way to build the much-needed mutual respect and trust among concerned stakeholders. Furthermore, development agencies or organisations should incorporate effective feedback mechanisms within their policy frameworks.

It is widely acknowledged that development practitioners have a lot to learn from their clientele system. Contrary to the patronising viewpoints of experts about grassroots communities as mostly unwilling to embrace change, the latter have a wealth of knowledge, which experts can use to devise better and enduring development agendas. As Robert Chambers rightly puts it, '…one first step is for outsider professionals, the bearers of modern scientific knowledge, to step down off their pedestals, and sit down, listen and learn' (Chambers, 1983, p. 101). Utlimately, this can be achieved through promoting approaches that strengthen the culture of co-creation of knowledge and solutions among development practitioners and their clientele.

Considering the precarious conditions in which development experts work most of the time, there is need to devise better ways to enhance their conditions of service and insure them against life threatening situations associated with their unique assignments in some ecologically and politically challenging terrains. This book offers a compelling reason why policy makers should explore new avenues of crafting and reforming policies. That said, the uniqueness and strength of this book mainly lie in its unorthodoxy and style of reaching out to interested readers, including people who currently work and will still work in the development-orientated organisations and institutions. The reflective and reflexive approaches used in compiling the stories are a deliberate attempt to help those keen in implementing development work differently now and in the future.

References

Chambers, R. (1983). *Rural development: Putting the last first* (pp. 1–218). Essex: Longman Scientific & Technical.

Grindle, M. S., & Thomas, J. W. (1991). *Public choices and policy change: The political economy of reform in developing countries* (pp. 2–17). The Johns Hopkins University Press.

Printed in the United States
by Baker & Taylor Publisher Services